The Chemistry of Wood Preservation

The Chemistry of Wood Preservation

Edited by
R. Thompson, CBE
Consultant

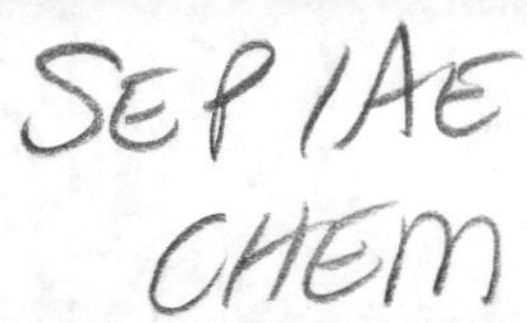

The Proceedings of a Symposium organised by the Industrial Division of the Royal Society of Chemistry held on 28 February - 1 March 1991 at the Laboratory of the Government Chemist, Teddington, UK

Special Publication No.98

ISBN 0-85186-476-7

A catalogue record for this book is available from the British Library

Published by The Royal Society of Chemistry,
Thomas Graham House, Science Park, Cambridge
CB4 4WF

Printed by Woolnough Bookbinders Ltd., Irthlingborough, Northamptonshire

Preface

The effective and lasting treatment of wood against insect and fungal attack grows in importance as some forestry reserves decline and as cost increases feed through to the building trade and other timber users. At the same time environmental pressures bear ever more heavily on the types of chemicals and processes employed in the preservation industry.

This book records the proceedings of an international meeting arranged to address such issues. The fifteen principal chapters are based upon papers presented by invited experts to a combined audience of preservation practitioners and non-specialists at a conference held in early 1991 under the joint auspices of the Society's Industrial Division and the Laboratory of the Government Chemist. The lecture programme was designed with this publication directly in mind, as no known single text source covers the interwoven chemical, biological, technological, environmental and legislative aspects of the subject.

Chapter sequence follows the logical pattern of the conference, beginning with a review of the biological threats to be contended. There follow historical through to state-of-the-art accounts of aqueous, organic solvent and non-liquid treatment processes. Preservatives increasingly must meet international product and environmental standards, which along with the related test, analytical and quality control procedures are described and referenced.

The wood preservation industry engages heavily in research, addressing a range of needs associated with cost, safety and performance efficacy, but not neglecting a search for a better understanding of the finer chemical mechanisms involved. Remaining problems are outlined in strategies for further research and development.

Not all aspects of the subject could be covered in a two-day programme, but it was considered that this book should for completeness include contributions on, inter alia, termite control (which is essentially a deterrent approach), the prevention of unsightly sapstain fungi (that deface rather than destroy wood) and vapour-phase introduction of boron.

The editor gratefully acknowledges the assistance of Professor G F Phillips OBE and Mr R Lees of the Laboratory of the Government Chemist in organising the conference, and expresses appreciation for the invaluable advice of Mr L T Arthur of Borax Research concerning programme content. The Society is grateful to the British Wood Preserving and Damp-proofing Association, the Institute of Wood Science and the International Research Group on Wood Preservation for publicising and supporting the meeting.

Raymond Thompson

Immediate Past-President, Industrial Division

Contents

Wood Preservation: The Biological Challenge

D.J. Dickinson

DEPARTMENT OF BIOLOGY, IMPERIAL COLLEGE OF SCIENCE
TECHNOLOGY AND MEDICINE, LONDON SW7 2BB,UK

1 INTRODUCTION

The biological problems associated with finding effective preservative systems for wood are often not fully appreciated by the chemist, in the same way the chemical properties and complexities of formulation technology are often little appreciated by the biologist. For the chemist it is not important to have anything but a passing knowledge of the names and taxonomy of the causal organisms but rather an appreciation of the environmental factors, the types of organisms with regard to their physiological grouping and the relationships between the two. This is essential if a close co-operation between the timber biologist and preservation chemist is to be achieved. In this overview the author will attempt to summarise the main groups of organisms, the environmental factors and the interactions between the two which are important in attempting to understand the complexity of "The Biological Challenge" that faces those who would attempt to preserve timber from the agents of decay.

The environment and ecology are very topical subjects but it should be remembered that the study of ecology is not new. Biologists have always considered the relationships between organisms and their environment. The study of micro-organisms in this context, or "microbial ecology" as it is called, is considered to be one of the oldest branches of microbiology. Microbiologists have always been fascinated by the organisms associated with specific habitats and the interactions between the organisms themselves and with the environment. Within wood, we are only just beginning to fully appreciate all the organisms involved and to understand the inter-relationships. The study of wood decay, its protection and the eventual fate of preservative treated wood has proved to be a superb example of applied microbial ecology.[1] Once the factors involved are understood by the producers of chemical protection systems, the easier it will become to utilise timber to its best advantage in a whole range of different environmental conditions.

In studying wood decay from an ecological point of view it is necessary to define a few basic principles and terms. Ecology is the study of the relationship between organisms themselves and the abiotic factors of the environment. This study may be restricted to a particular "habitat" which in our case is the "polymeric, cellular ligno-cellulosic composite" we call wood. The condition in which wood is used will vary, i.e. the environmental factors will vary. Under any set of environmental factors the organisms that will be present will be selected by those conditions. They will be ones most "fit" for the prevailing conditions, that is, the environmental factors select the "characteristic community". Different environmental conditions will give rise to different, if somewhat overlapping, communities. The term we use to describe any characteristic community, set of environmental factors and the non-biological surroundings is an "ecosystem". It follows that if the prevailing conditions in the environment differ in different end use situations for timber, that different ecosystems will exist, each with its characteristic community. The biologist and chemist need to adjust the preservative requirements to each of the different ecosystems in order to protect wood. Such an approach to understanding decay in use will lead to preservatives specifically designed for particular end uses where not only the physical compatibility aspects are taken into account but the demands of the specific ecosystems are also met.

2 THE DIFFERENT ECOSYSTEMS IN THE FOREST

Before we can consider the specific ecosystems that exist within timber in constructions we need to consider the problems that exist prior to utilisation of timber in structures. In the forest and storage situation we tend to get a much more dynamic situation where the agents of decay act very quickly and the different environmental conditions tend to merge. This is in contrast to structures where the prevailing conditions tend to be much more distinguishable. Prior to utilisation we can consider several different situations: The standing tree; felled and green sawn timber; wood in storage.

2.1 Standing Trees. In the standing tree, the situation differs to that in timber in that the tree is living and growing. What we have is a true host/parasite relationship where the living tree may be able to respond to invasion by micro-organisms. Once timber is felled these host responses are lost and many more sites of entry for organisms are opened up. In trees the most important forms of decay are the so called butt or heart rots. Normally, fungi enter the tree via wounds above ground or through the root system. The sapwood of standing trees is normally free of decay due to a combination of host response and, very importantly, due to the fact that it is too wet for decay fungi to grow. Most decay is therefore limited to the heartwood which is drier and lacks the hosts responses. Decay in standing trees is often not evident until the tree is felled or blown over due to weakening of the stem. In the U.K. the most important fungi causing decay in standing trees are *Heterobasidium annosum* followed by *Armillaria mellia*. Although the invasion and control of decay in forest trees is of great interest and importance to the forest pathologist, it is of little interest to the preservation chemist. As a rule, timber which is to be utilised in structures should be free of all pre-infections. In certain circumstances infection can be carried over into the sawn timber and may continue to develop in structures. In such situations, particularly with large dimensional timbers, it is normal to rely on kilning and heat sterilisation to ensure that the timber is free of any such potential problems.

2.2 Felled Timber. Once the tree is felled the situation is quite different. The cut ends of the logs and wounds from branch removal open up numerous sites for the entry of micro-organisms. In the early stages, the sapwood will be too wet for most fungi but, unless preventative measures are taken, it will soon dry out to a level which makes it highly susceptible to invasion by a whole range of fungi.

Fresh timber contains large amounts of stored sugars and starches in the ray tissues. This food source is readily available to many organisms unlike the cell wall polymers which can only be attacked by fungi which have the enzymes capable of degrading the complex chemical structure of the wall. Such fungi are the true wood decay organisms. The simple, stored carbohydrates therefore represent a readily available source of food for micro-organisms and are always the first to be utilised.

As a general rule, when food is readily available and other factors such as water and oxygen are not limiting, the range of organisms or species diversity will be great. In extreme environments where food is scarce or some other factor is limiting or toxins are present, then species diversity is often low. In fresh timber, with all its stored food as simple carbohydrates, many fungal species are able to grow in it, but those which can grow fastest tend to predominate. Initially this will be the micro-fungi such as the moulds and "blue-stain" fungi. As this readily available food source is used up then those fungi that can attack the cell wall directly to derive food will tend to take over. In a felled tree or saw-log the sapwood may still be very wet and it is the white-rot basidiomycete fungi which prefer, or more correctly can tolerate, the wetter conditions that are most common under these conditions. As the log dries out, checks develop and the inner, drier heartwood is exposed. At this time highly destructive brown-rot fungi will invade the log. The micromorphological features of these two important groups of wood rotting basidiomycetes are described later.

In the forest therefore on a felled log we see a fairly rapid succession of organisms that invade the log and, within a short time, decay becomes established.[2] In order to prevent colonisation of logs, it is important to fell trees in the cold season and convert them to dry timber as quickly as possible. If delay cannot be avoided or storage is desirable for strategic reasons, it is necessary to store them in such a way as to prevent fungal invasion.

2.3 Wood in Storage. If the logs are prevented from drying out by storing in ponds or under water sprays, fungal invasion is completely prevented. These conditions are very similar to those present in the standing tree, except that the water is static within the log. Under such conditions, certain bacteria can grow but generally they do not present a major problem providing a good turn round of logs is maintained. In summary, we can say that the control of fungi in sawn logs is one of logistics and that the sawmill should only be converting timber free of infection to give clean infection free timber.

2.4 Converted green timber. At this stage, sawn timber is once again at a high risk from infection as in the green log. The wood is wet, full of simple carbohydrates as a food source and the freshly cut surfaces present no defence to infection. Unless the wood is rapidly dried the same fungi that infect green logs will rapidly invade the exposed wood. The development of stain and mould in sawn softwoods varies between wood species, but generally speaking all sapwood is susceptible with time, necessitating preventative measures. Any chemical treatment designed to prevent infection will have to have a very wide spectrum of activity. If the most prevalent organisms are controlled, other resistant fungi will soon take advantage of the new situation. The challenge to the chemical industry is very attractive but success is often difficult to achieve.[3,4] Cocktails and attention to presentation of the chemical are very important factors in achieving good protection.

Once dry, the timber is no longer susceptible to attack by micro-organisms but can still be attacked by insects. In timber yards, particularly with hardwoods, *Lyctus* beetles need to be controlled. These insects have been compared to the blue stain fungi in that they feed on the stored starch in the rays. In obtaining this the larvae of the beetles totally destroy the sapwood, reducing it to a powder, hence the common name, powder post beetles. These beetles make it necessary to treat most seasoning hardwoods with effective insecticides which are usually incorporated with the antistain preparations.

In the forest and timber yard situation, it can be seen that the conditions are rapidly changing, particularly with respect to moisture content. Consequently, the characteristic organisms are also in a progressive state of change, showing a colonisation sequence from the bacteria, through the micro-fungi to the white and brown rot fungi. In buildings, the environmental conditions tend to be much more stable and the individual ecosystems, controlled primarily by moisture content, are more distinguishable. Within Europe we have defined these different ecosystems as "Hazard Classes". These classes are essentially moisture regimes and as such each select characteristic communities of wood inhabiting organisms. These ecosystems or hazard classes therefore have defined environmental conditions, characteristic communities, which in turn define the necessary preservation requirements.

3 EUROPEAN HAZARD CLASSES.

The basis and importance of these classes are covered elsewhere in this publication by Brooks. They are defined in terms of service conditions with reference to the generalised moisture content and the prevailing biological agents of deterioration. Although moisture is the all important factor, temperature due to geographical location is also important throughout Europe, particularly with regard to the occurrence of certain insect pests. As a result, occurrence within Europe of each biological agency is indicated according to whether it is universally present or restricted to certain geographical locations. Table 1 lists five such "Hazard Classes" which are defined within constructions in Europe.

Table 1. European Hazard Classes

CLASS	SITUATION	WETTING	WOOD MOISTURE
1	Above ground covered	Permanently dry	<18%
2	Above ground covered- risk of wetting	Occasional wetting	Occasionally >18%
3	Above ground uncovered	Frequent wetting	Often >20%
4	In the ground or fresh water	Permanent exposure to wetting	Permanently >20%
5	In salt water	Permanent exposure to wetting by salt water	Permanently >20%

<u>3.1 Hazard Class 1.</u> In hazard class 1 the wood is above ground and protected therefore permanently below 18% moisture content. Under these conditions micro-organisms are unable to grow due to the absence of free water. However, wood destroying beetles and termites

are able to survive and attack the wood, but this hazard will vary according to prevailing temperature in the different regions of Europe.

Three types of wood boring beetle are considered important in Europe. They are *Anobium punctatum*, the common woodworm or furniture beetle, *Hylotrupes bajulus*, the house longhorn beetle, and *Lyctus brunneus*, the powder post beetle. Termites exist only in the most Southern regions of Europe but present a very serious problem where they occur. In the tropics they represent the most severe hazard to building timbers.

The wood boring beetles show a complete metamorphosis, i.e. in their life cycle they show a change of life form after a pupal resting stage. The eggs are laid on rough surfaces and hatch into small larvae which feed and grow in the timber. This is the destructive stage of the life cycle. This feeding stage may last for up to several years with *Anobium punctatum*, depending on the amount of body building nitrogen contained in the timber. Eventually the larvae pupate into a resting stage from which the sexually mature adult eventually emerges to mate and infect more timber with eggs to complete the life cycle.

Anobium punctatum is by far the most common wood borer in the U.K. With the advent of central heating, the wood moisture content in our homes has been lowered and this pest is likely to decline under these conditions as most homes are near the limit for its survival. In continental Europe the problem of *Anobium* pales into insignificance compared to *Hylotrupes bajulus* which is far more destructive, and is capable of causing failure in structural timbers when the sapwood content is high.

Termites exist throughout the tropics but extend into Southern Europe. The termites belong to the Insect group known as the Isoptera. Their life style is quite different from the wood boring beetles. Two distinct groups of termites attack wood. The Subterranean Termites build their nests in contact with the ground and are responsible for the characteristic termite mounds. They forage over a distance for their food and build tunnels between their nests and the source of timber. Control of subterranean termites can therefore consist of excluding them from buildings by such means as toxic barriers and soil poisoning or by direct treatment of the timber.

Dry Wood Termites, in contrast, live independent of the ground excavating clean galleries within the timber structure. The colonies are much smaller but once established in a building can be difficult to eradicate. Prevention by exclusion of flying insects and conventional wood preservation are the most effective measures.

3.2 Hazard Class II. In Hazard Class II the timber is defined as being above ground, under cover, but in danger of some wetting. The moisture content of the timber will be above 20% for periods of time. Under such conditions the insect hazard will still be present and in fact is likely to be more severe, because class I is really a marginal situation for wood borers. The important factor is that under these environmental conditions there is sufficient moisture for some of the fungi to develop. Only those fungi that can grow quickly, such as the moulds which take advantage of conditions where water condensation occurs, and the decay fungi able to grow in fluctuating, drying conditions will generally be present. The brown-rot fungi, which are the last to occur on logs in the forest are the decay fungi most common in Hazard Class II. Brown-rot in buildings, caused by accidental wetting due to bad maintenance is best exemplified by "Wet" and "Dry" rot, caused respectively by a range of fungi, but predominantly *Coniophora puteana* and by *Serpula lacrymans*, the true dry-rot fungus.

The brown-rot fungi are capable of degrading the wood cell wall itself. They remove the cellulose and hemicellulose component but leave the lignin behind essentially unaltered.[5] The hyphae penetrate the lumen of the cell and lie on the inner surface of the cell wall. The appearance of both the hyphae and the S3 layer of the wall change very little but the other layers are reduced to a brown amorphous mass.[6] The brown-rot fungi are capable of destroying the wall at a distance from the hyphae, probably involving a chemical modification of the wall by a mediator such as oxalic acid to allow the cellulolytic enzyme to attack the lignin encrusted cellulose. Much work and speculation is currently in progress to establish the precise mechanisms of cell wall degrade by the brown-rot fungi.[7]

<u>3.3 Hazard Class III.</u> In Hazard Class III the timber is above ground, not covered and subject to regular wetting. The timber may be above 20% moisture content for long periods of time. Under these conditions the insects and brown-rots will continue to be a problem but timber at a higher moisture content for longer periods will be susceptible to attack by the white-rot Basidiomycetes. The classic example of such conditions is found in external, painted joinery or fencing rails. This hazard class probably represents the optimum conditions for decay as found in the green, drying log in the forest. All the agents for decay can survive and the climax, wood rotting Basidiomycetes are unhindered by the wood either being too dry or too wet. Major ecological studies of simulated window joinery have shown the same ecological sequence as occurs in the forest.[8] The bacteria invade first, giving way to the micro-fungi and eventually leading to the rapid decay of the timber by the white and brown-rot fungi.

The hyphae of the white-rot fungi actively penetrate into the cell wall and lie along the lumen surface.[9] Lysis of the wall occurs along the line of the hyphal contact, forming a groove or trough with a central ridge on which the hypha rests.[6] As the hyphae branch, the grooves coalesce eroding the cell wall. Both the lignin and cellulose present in wood are degraded and this can result in weight losses in excess of 90%.

Exposed timber in Hazard Class III also represents a specific ecosystem at the surface of the timber or at the interface of the wood and applied decorative finishes. This is a particularly harsh environment and only certain specialised fungi can grow, particularly *Aureobasidium pullulans* which causes the so-called blue-stain in-service; the term used to describe these growths on exposed wood. Although these fungi can utilise available stored food, at the surface of wood they probably depend on the partial physico-chemical breakdown of the wood to allow access to the cellulose as a food source.[10]

<u>3.4 Hazard Class IV.</u> In Hazard Class IV the timber is in ground contact or in fresh water and therefore permanently exposed to wetting, resulting in

the timber being above 20% most of the time, and often very much higher. Under such conditions all the organisms present in the other hazard classes will be present. However, for much of the time the wood will be too wet for the wood rotting Basidiomycetes. Under such conditions, a group of decay fungi known as the soft-rots are prevalent. They are highly tolerant of the wetter conditions but are normally displaced by the true wood rotters when conditions favour the Basidiomycetes.

In ground contact the conditions will vary from near anaerobic conditions well below the ground where decay will not occur, to drier conditions suitable for brown and white-rot fungi in what are essentially Class III conditions. In the wetter wood fungi imperfecti and ascomycetes can grow and cause the characteristic soft-rot associated with Hazard Class IV.

These are grouped together on the basis of forming cavities in the S2 layer of the wood cell wall which is destroyed by the formation of such cavities. Savory showed that soft-rot grew in S2 layers of the secondary cell wall, forming chains of cavities with pointed ends that were orientated parallel to the cellulose microfibrils of the cell wall.[11,12] Corbett showed how the fungal hyphae penetrated the S2 layer of the wood cell wall.[13] A short side branch of a hypha lying in the lumen penetrated the S3 layer perpendicular to the cell wall and formed at "T" branch. A cavity was eroded around the hyphal branch cross of the "T" branch and the hyphae increased in girth. Fine hyphal strands, termed proboscis hyphae formed the initial extension of growth in the wall which then increased in girth forming a new cavity around them.[13,14,15] Much speculation still exists as to the initiation of "T" branching and the characteristic diamond shape of newly formed cavities. In systematic isolation studies of wood in ground contact the soft-rot fungi are the first decay fungi to be isolated. In the drier wood they are displaced by the brown and white-rots but in the wetter wood, or in other conditions such as in the presence of toxins, basidiomycete growth is inhibited and the soft rots assume the climax position with regard to decay.

<u>3.5 Hazard Class V.</u> Hazard Class V consists of timber in the marine environment, permanently exposed to wetting by salt water. In practical terms much of the timber in a marine structure will be in Hazard

Class III, i.e. above ground, uncovered. In the sea water itself the timber is in a very severe situation. Decay by Basidiomycetes does not occur but marine soft-rot fungi are very common causing surface softening of the timber, but the principal agents of decay in this situation are the Marine Borers.

In European waters the most common marine borers are Shipworm (*Teredo navalis*) and Gribble (*Limnoria lignorum*). Shipworm or *Teredo* is a bivalved mollusc related to the sea snails and mussels. It is a soft, worm-like animal with its shell modified into hard grinding jaws. The larvae are part of the microscopic zooplankton and swim freely in the sea until they settle on timber. They develop a shell with which they bore into the wood and lodge there, growing into large worms in holes up to 5mm in diameter. They destroy the wood by making a massive network of such holes throughout the timber. Gribble is a small shrimp-like crustacean about 4 mm in length. It bores into the surface of the wood and lodges near the surface making numerous side burrows. The combination of this boring and wave action causes rapid erosion of marine timbers.

Timber in constructions can therefore be seen to represent 5 different ecosystems, each with its characteristic environmental conditions which has the effect of selecting a characteristic community of wood destroying organisms which tend to mirror the natural situation in the forest.

4 TREATED TIMBER IN SERVICE

Once treated with chemicals, i.e. a wood preservative, the timber represents a totally new habitat. An additional, abiotic factor has been added to the system, i.e. a toxin. Whereas the water availability tends to be the limiting factor for decay to occur in untreated wood, the toxins themselves become the all important abiotic factor determining the types of organisms that can live in the wet wood.

Wet treated timber will eventually succumb to decay by micro-organisms, normally towards the end of its predicted, desired service life. Treated wood , from an ecological stand-point, must be viewed as a harsh environment for micro-organisms. Under such conditions few organisms can grow, but it is an accepted concept of microbial ecology that eventually an organism capable of exploiting the resource will eventually do so. If no organism in the immediate vicinity can live in the treated wood, others capable of doing so will eventually be transported to it, or the organisms themselves will change to exploit the substrate. In other words, as stated, all other things being suitable for decay, treated wood will eventually decay given sufficient time. The mechanisms by which this occurs is of great interest to the preservation mycologist not only from the academic point of view but it enables us to predict the long term performance of new systems and to eliminate those preservatives which may fail early to otherwise unexpected factors.

There are several very good examples in current preservative systems which illustrate well the toxic habitat and how organisms overcome it to utilise the wood as a food source. Probably the best understood situation is the selection of preservative resistant basidiomycetes in service. A good example is the failure of the Copper Chrome preservatives to the copper tolerant brown-rot fungi, such as *Poria* species. The copper fungicide is precipitated as the non toxic insoluble oxalate in the wood allowing the fungi to decay the wood. This has led to widespread inclusion of arsenic and borates in copper chrome preservatives, specifically to control this problem.

With the introduction of new organic preservatives, it is very important to realise that some basidiomycetes are capable of degrading them. The white-rot fungi probably attack such fungicides in the same way as they break down the phenolic polymer lignin. Probably the oldest example of the selection of a tolerant basidiomycete by preservative treated wood is the decay of poorly creosoted transmission poles by *Lentinus lepideus*. This fungus is by far and away the most common fungus causing decay in the centre of creosoted commodities, where the conditions are unsuitable for other fungi. The phenomena of tolerance by the decay fungi themselves has been well appreciated by the preservation industry for many years. As a result, tolerant isolates of decay basidiomycetes are always included in any testing, development programme for any new wood preservative.

Other modes of failure of preservative wood have also begun to be more fully understood in recent years. Often there is a shift in the physiological group of fungi that normally cause decay. This is particularly the case in high hazard situations where all the fungi are capable of growing. By controlling one group, such as the Basidiomycetes, a less competitive one can grow that is tolerant to the preservative. A classic example of such a shift is in the failure of copper chrome arsenate treated hardwoods. Although decay by basidiomycetes is clearly controlled, preservative tolerant soft-rot fungi are able to decay the wood.[16] Several factors are at work but the over-riding factor is the presence of the toxins in the wood. Even in treated softwoods where both the basidiomycetes and soft-rot fungi are controlled, under certain conditions cell wall degrading bacteria can decay the wood and eventually let in other organisms by causing changes in the toxic environment.[17]

In some cases, all the organisms capable of causing decay may be controlled, but other non-decaying microfungi may be able to grow and modify the toxin so that the decay fungi can then come in later and decay the wood. This modification of a toxic environment by one group of organisms which then allow a second group to grow is quite common in microbiological systems and could well be a very important factor with new preservative systems, particularly those based on organic chemicals. Carey in her ecological studies of L joints clearly showed that several of the tolerant non-decaying early colonisers were capable of detoxifying Tributyl-Tin-oxide finally allowing the white-rot fungi to establish in the joints.[8,18] It is now generally accepted that the failures of timber treated with Quaternary Ammonium compounds in New Zealand were probably due to detoxification by staining micro-fungi which then allowed decay to proceed. Most failures of treated wood take a very long time and in fact represent the life of the commodity in service. It is very important to understand these systems of failure in treated wood in order to prevent premature failure of new systems.

5 CONCLUSIONS

The decay of timber must be regarded from a dynamic, microbial, ecological point of view. All wood, even treated wood, will fail to the micro-organisms eventually. Wood in structures can be regarded as a series of ecosystems with its characteristic communities. This makes it possible to select and design preservatives for specific end uses.

Where water is not limiting, it must be remembered that toxins may simply select out another group of organisms. This is particularly important in ground contact situations. Until we fully understand and can mimic these situations in the laboratory, it is essential that preservative systems for ground contact are allowed to be stressed against the whole complement of micro-organisms, even though they are not normally considered to be true decay fungi.

REFERENCES

1. D. J. Dickinson and J. F. Levy, In 'Methods in Microbiology', Academic Press, London. (Editors, R. Grigorovora and J. R. Norris). 22, pp479-496. 1990.

2. D. J. Dickinson and R. J. Murphy, In the records of 'Annual Convention of the British Wood Preservers Association'. 1991. (In Press).

3. D. J. Dickinson, Holzforschung, 1977, 31 (4), pp121-125.

4. D. J. Dickinson, In Proc. I.U.F.R.O. Wood Protection Subject Group. 1987. pp350-361.

5. J. F. Levy, In 'Celluloses, Sources and Exploitation', Ellis Horwood Ltd., London. (Editors, J. F. Kennedy, G. O. Phillips and P. A. Williams). Chapter 50, pp397-407. 1990.

6. A. F.Bravery, J. Inst. Wood Sci., 1971, 3 (30), pp13-19.

7. P. J. Harvey and J. M. Palmer, Spectrum, British Science News. 217 (1). 1989.

8. J. K. Carey, In 'Biodeterioration', Wiley, New York and London. (Editors, T. A. Oxley and S. Barry). 5, pp13-25. 1983.

9. J. F. Levy, In 'Technology in the 1990's: Utilisation of Lignocellulosic Wastes'. Phil. Trans. Royal Society, Series A, 321, pp423-433. 1987.

10. D. J. Dickinson, In the records of 'Annual Convention of the British Wood Preservers Association'. 1971.
11. J. G. Savory, Ann. App. Biol., 1954, 41, pp336-347.

12. J. G. Savory, In the records of 'Annual Convention of the British Wood Preservers Association'. pp3-35. 1955.

13. N. H. Corbett, PhD Thesis, University of London. 1963.

14. N. H. Corbett, J. Inst. Wood Sci., 1965, 3 (14), pp18-29.

15. A.Crossley and J. F. Levy, J. Inst. Wood Sci., 1977, 7 (42), pp30-33.

16. D. J. Dickinson, Mat. und Org., 1974, 9, 21-33.

17. T. Nilsson and G. F. Daniel, In 'Proc. 16th Conv. Deutsche Geselschaft fur Holzforschung. Munster. 1982.

18. J. K. Carey, PhD Thesis, University of London. 1980.

Industrial Wood Preservatives — The History, Development, Use, Advantages, and Future Trends

M. Connell

HICKSON TIMBER PRODUCTS LIMITED, CASTLEFORD, WEST YORKSHIRE WF10 2JT, UK

1 GENERAL PRINCIPLES

No timber is immune to deterioration if exposed for a sufficiently long period to the natural environment. The serviceable life of individual pieces varies considerably, depending on the species concerned, the amount of sapwood present, the use to which the timber is put and the situations and environmental conditions to which it is exposed.

The living organisms which are natural enemies of wood generally use it as a source of food and shelter or as a place of incubation for their young. In order to survive the organism must have a source of food, moisture, and air at the right temperature. If any of these factors can be removed then the risk of attack by these agents will be eliminated. For example, the untreated wood components from the camps set up by Captain Scott on his expedition to the South Pole were found to be intact upon being rediscovered by Dr. Fuchs, 50 years later. The Egyptians were aware that timber would not decay if kept dry and untreated sycamore coffins have been found in Pharaoh's tombs undamaged by insects or fungi after 4000 years. Further examples include intact iron-age canoes found immersed in marshes in Denmark and remains of the Marie Rose recovered from waters off the South coast of England where lack of fungal decay was due to the unavailability of oxygen.

The problem is that controlling temperature, oxygen and moisture content can be extremely difficult and the only way to prevent the growth of degrading organisms is to limit the supply of food. In the case of wood this can be done by treatment with chemicals which are toxic to these organisms and so making it unattractive as a food source.

2 HISTORY AND DEVELOPMENT OF TIMBER PRESERVATION

The modern industrial wood preservation industry did not begin until the early 19th Century but chemical wood preservation is not new. Although timber was in plentiful supply in ancient times and generally thought of as being unworthy of protection certain objects were subjected to preservative treatments. Pliny noted that the statue of Diana at Ephesus, was saturated with oil of nard through small holes bored in it. Both the Romans and the Greeks used oil, tars and resins, extracted[1] from resistant timbers to preserve structures such as bridges.

The use of inorganic chemicals can be traced back to the Egyptians, who preserved their rulers by mummification which involved steeping the body in Natrum, a mixture of various sodium salts for seventy days and then in an oily or bitumous substance for a similar time. The Chinese, 2000 years ago, immersed their wood in sea water or the water of salt lakes prior to using it as a building material. Well preserved props have recently been removed from Roman mines in Cyprus and examination has shown them to contain metallic copper, probably deposited there electrolytically from copper-rich soil in the area.

Probably the earliest person recorded as using a wood preservative was Noah, who when building the Ark was instructed by God to "pitch it within and without with pitch". A further reference in the Bible to wood degradation and possible methods of treatment occurs in the Old Testament Book of Leviticus which describes in detail an attack of dry rot or the "fretting leprosy of the house".

Following the decline of the Roman Empire little happened in the development of timber preservation until the end of the 16th Century. The major trading countries in Europe relied heavily on their navies and it was the enormous cost of damage to timber in the sea which led to a more concerted effort to identify more effective wood preservatives.

Although many materials were tried the results were generally disappointing. In the early 19th Century the Royal Navy was involved in wars against Napoleon and in North America and the problems of decay and insect attack coupled with a general shortage of timber caused the Admiralty to look at chemical wood preservation more seriously. In 1812 an experiment to inject resinous vapours into timber ended in disaster when the apparatus exploded with fatal consequences for the workmen.

Different chemical formulations were tried during the 19th Century including mercuric chloride, copper sulphate and zinc chloride but all had the drawback of being soluble in water and *therefore of little use in external situations.*

During this period the intense pace of the industrial revolution saw the development of the steam engine and the use of coal as a primary fuel. The rapid growth of the railways and telecommunications system was requiring large quantities of sleepers and poles.

A by-product of coke production was creosote oil which was plentiful and cheap. In 1838 the modern industrial timber preservation industry was born with the invention of pressure impregnation of timber by Bethell. The Bethell process involved full cell penetration and was complimented by other developments including empty cell processes by Rueping and Lowry and sap displacement processes by Boucherie and Boulton.

The treatment of sleepers and poles applying creosote via pressure impregnation grew rapidly both within the UK and other European countries. In 1900 the Director-General of the Silesian Kattowitz mines approached Dr. K.H. Wolman to consider improvements to existing methods of preservation and in 1903 the largest creosote pressure cylinder in the world was installed for the treatment of pit props, telegraph poles and railway sleepers.[2]

At this time the fungicidal properties of hydrogen fluoride and its salts were beginning to attract attention and Wolman directed his attentions to developing a wood preservative based on fluoride. His work met with some success and in 1907 he patented his first fluoride containing preservative (Austrian Patent No. 31180). The preservative suffered the same deficiencies as single salt type preservatives introduced at that time, essentially a narrow spectrum of activity and lack of permanence in the treated timber.

Further development by Dr. Wolman resulted in the first multi-salt preservative of the TRIOLITH type (German Patent No.299411) with a composition of:-

Sodium fluoride	*85%*
Dinitrophenol	*10%*
Sodium dichromate	*5%*

The preservative appeared at the time to be a major break through and its use was boosted during the 1914-18 war when creosote was scarce. The German State Railway turned to TRIOLITH for the treatment of their sleepers and it was also exported to Switzerland, Denmark and Sweden. In order to improve the termicidal properties of this type of preservative, sodium

Sodium fluoride	60%
Sodium arsenate	20%
Sodium dichromate	15%
Dinitrophenol	5%

Dr. Wolman succeeded in finding a market for his preservatives in the USA, Africa and Asia although he was not permitted to export the preservative into the States. In 1923 he negotiated a Licensee agreement with the American Lumber and Treatment Company of Chicago and New York (now Hickson Corporation). This company expanded rapidly and in the 1930's its consumption of preservatives surpassed many times that of Dr. Wolman's organisation in Germany.

The big drawback of these products was that the components had poor permanence in the treated timber and leached out in wet conditions. Dr. Wolman continued to search for ways to improve the permanence of the active components and discovered that by increasing the chromium content, the permanence of the arsenic component and to a lesser extent the fluoride content could be improved. Between 1922 and 1930 Dr. Wolman developed a number of preservative systems which he patented. Not one of these systems is now used commercially.

From a relatively early period in the development of these water-based preservatives the mixtures were known collectively as 'Wolman Salts'.

In 1933 the first copper, chrome, arsenate formulation (CCA) was invented by an Indian Government research worker, Dr. Sonti Kamesan in the form of copper sulphate, potassium dichromate and arsenic pentoxide. The product was called 'Ascu' and great interest developed during the 1930's culminating in the first full scale commercial installation of CCA-treated PTT poles in 1940-42 by the Bell Telephone system. The poles were exposed to various natural hazards and conditions and were thoroughly evaluated over the following 10 years. The results were so good that in 1953, the American Wood Preservers' Association (AWPA) included the Kamesan formulation as CCA-Type A, together with Boliden K33 as CCA-Type B in their specifications. These formulations are respectively high chromium:low arsenic and low chromium:high arsenic types.

During the early 1950's chemists and biologists at Hickson Timber Impregnation (GB) Ltd. were making extensive laboratory trials and evaluating the results of the Bell Telephone Systems' work with the resultant introduction of a 'Type-C' preservative. This Type-C CCA contains median arsenic:chromium proportions and the main debate during the 1960's centred on the optimum balance to achieve best fixation of the biocidal copper and arsenic *components.*

A number of investigations demonstrated the improved performance of the Type C product when compared to other CCA types.[3-8] *More recently this has been graphically demonstrated*[9] *in an EN84 leaching test comparison between CCA Oxides Types B and C (see graphs 1, 2 and 3).*

In 1966, BS4072 was first published. It specifies the two CCA's commercially available in Great Britain, Type 1 and 2. In 1969 the AWPA added a third product to its standards as CCA-Type C. The five CCA's may be conveniently compared by expressing the formulations as metal oxides although commercial products use salts and/or oxides as raw materials (Table 1).

Over the same period in which the use of CCA's has developed and expanded, other waterborne preservative products have been tried and proposed. The AWPA standards for example include four other products; acid copper chromate (ACC), ammoniacal copper arsenate (ACA), chromated zinc chloride (CZC), and fluorochrome arsenate phenol (FCAP). However, there is little demand for these products in today's market-place.

In the Germanic countries copper, chrome, boron (CCB) and copper, chrome, fluoride (CFK) products have become popular but have never succeeded to the same extent as CCA's. Comparative leaching tests carried out by Deon[10] *show CCA to be superior to copper, chrome, fluoride; copper, chrome, boron; and fluor, chrome, boron. The tests show clearly that the boron and fluoride components have poor fixation properties no matter how they are formulated.*

Today it is estimated that the annual world-wide sales of CCA products is in excess of 100,000 tonnes per year of which greater than 90% is based on Type C. In contrast the sales of CCB and CFK preservatives total some 10-12,500 tonnes per year.

A problem of more recent times is the significant increase in the incidence of decay in untreated joinery from the early 1960's.

Table 1 Copper-Chrome-Arsenates

Standard		Product	CrO_3	CuO	As_2O_5
BS4072 (Type 1)	e.g.	CELCURE A	46	17	37
BS4072 (Type 2)	e.g.	TANALITH C	52	19	29
AWPA (CCA-A)	e.g.	ASCU	65	18	17
AWPA (CCA-B)	e.g.	K33	35	20	45
AWPA (CCA-C)	e.g.	WOLMAN CCA	47.5	18.5	34

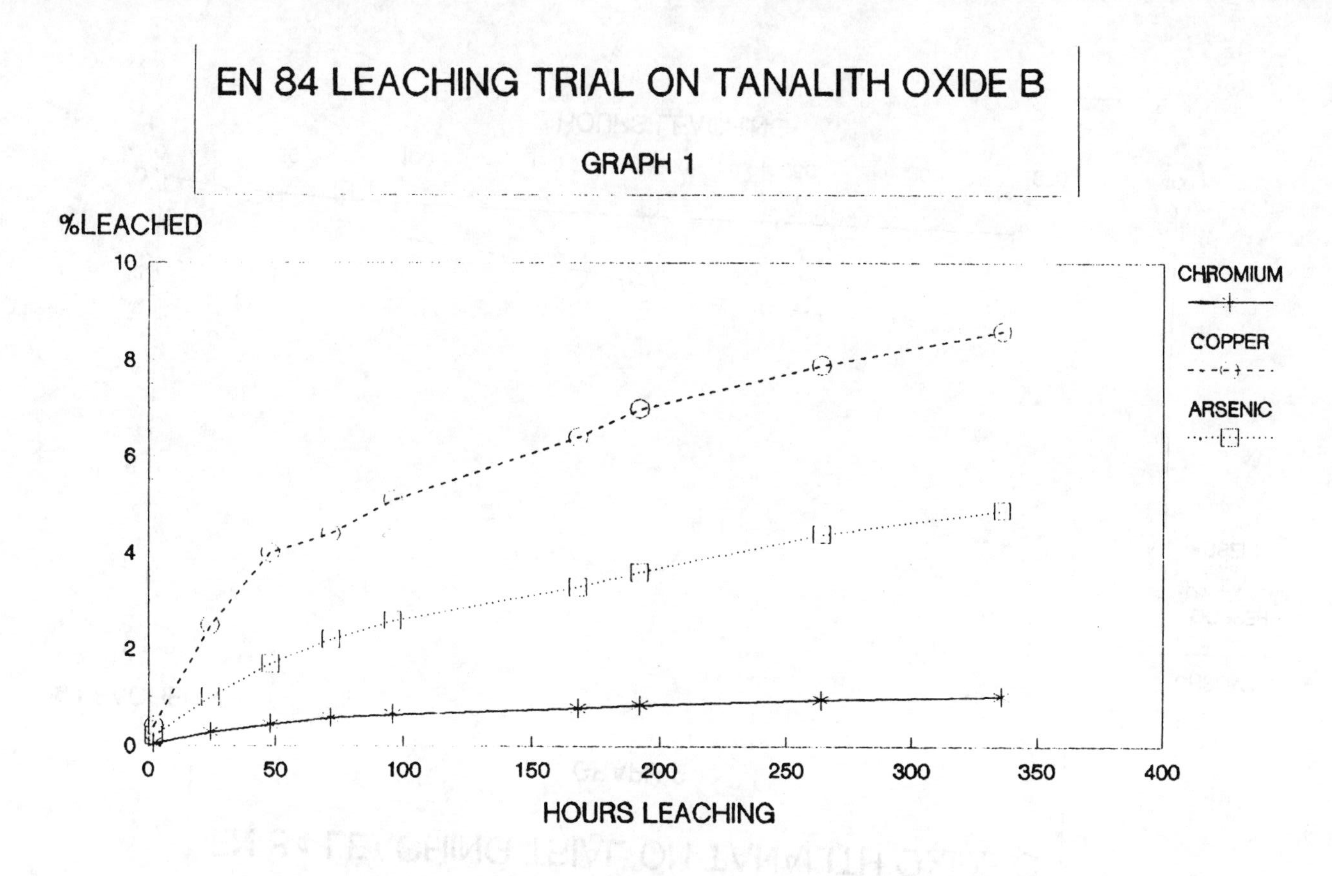
EN 84 LEACHING TRIAL ON TANALITH OXIDE B
GRAPH 1
%LEACHED
10
8
6
4
2
0
0
50
100
150
200
250
300
350
400
HOURS LEACHING
CHROMIUM
COPPER
ARSENIC

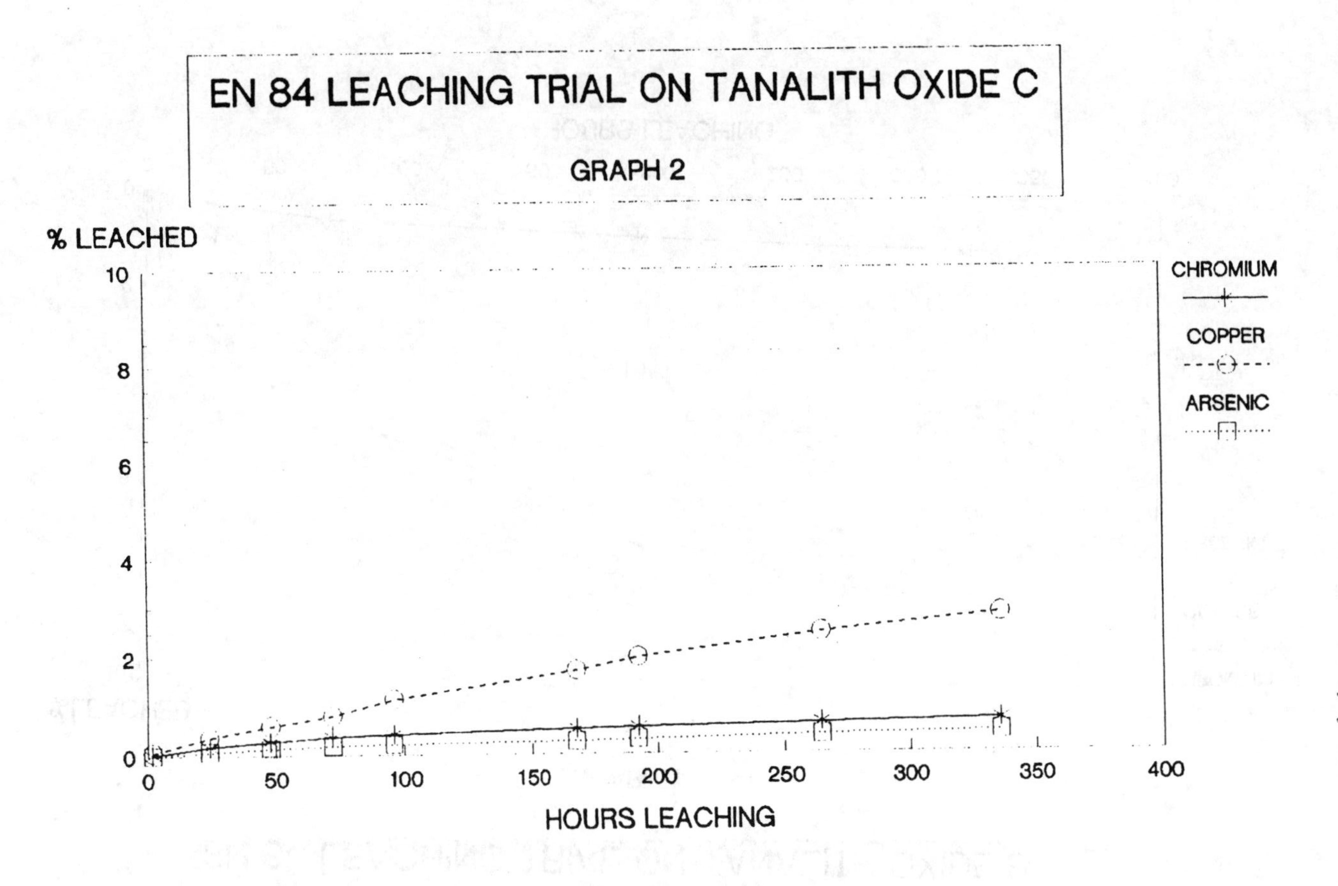
EN 84 LEACHING TRIAL ON TANALITH OXIDE C
GRAPH 2
% LEACHED
10
8
6
4
2
0
0
50
100
150
200
250
300
350
400
HOURS LEACHING
CHROMIUM
COPPER
ARSENIC

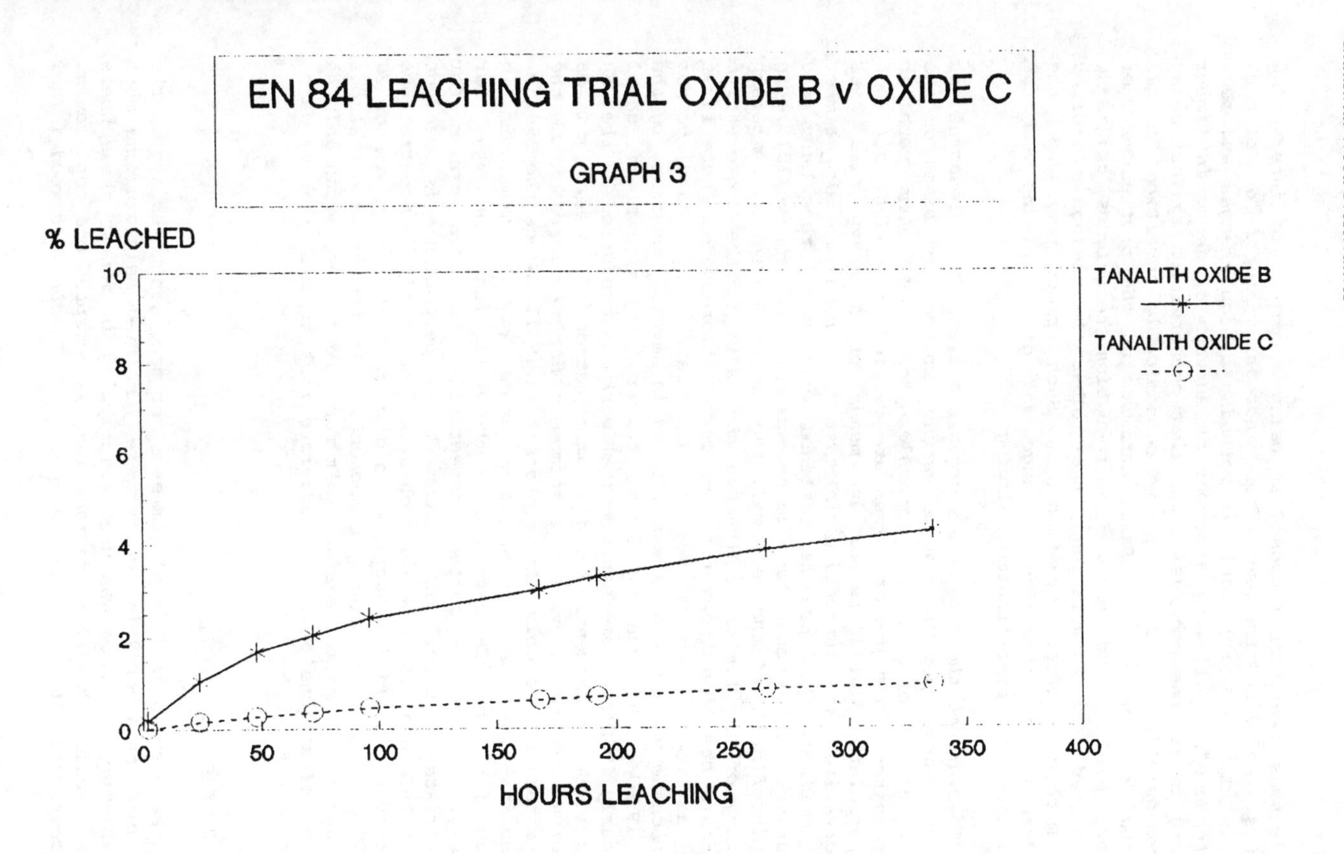

EN 84 LEACHING TRIAL OXIDE B v OXIDE C
GRAPH 3
% LEACHED
10
8
6
4
2
0
0
50
100
150
200
250
300
350
400
HOURS LEACHING
TANALITH OXIDE B
TANALITH OXIDE C

The reasons have been proposed as being a complex of interacting factors including the use of an increasing proportion of non-durable sapwood, constructional methods, design changes and poor maintenance". Following a number of surveys of decay in timber joinery in use one conclusion was that preservation against decay of non-durable timber species is economically justified for all external components.[12] Early attempts at the pretreatment of joinery used vacuum-pressure impregnation techniques utilising water-based FCAP formulations, the timber then being kiln dried and machined after treatment. Such processing has the disadvantage of dimensional change and grain raising and the requirement for post-treatment drying.

To overcome these problems organic solvent (OS) preservatives have been developed for double vacuum impregnation based on the historical performance of pentachlorophenol (PCP) and metallic naphthenates. The preservatives are carried in a light oil such as petroleum distillate or by means of a heavy oil. OS preservatives are attractive in that they penetrate dry timber very readily, they can be combined with water repellants and colourants, and because they do not combine with the cellulose in the microfibrils of the cell wall they do not cause any temporary swelling of the timber. Although the value of the preservative chemicals has been known for some time (chlorophenols were first used as wood preservatives in the late 1920's and copper naphthenate was quite well established in several countries by the late 1930's) they did not become commercially significant until the early 1950's. Some experimental work was conducted in the USA using vacuum treatments for the impregnation of joinery with OS preservatives but a viable system was not developed until the production of a double vacuum process in 1961. Since the 1960's, organotins such as tributyltin oxide (TBTO) sometimes in conjunction with PCP have become the major fungicides used for joinery. Gamma-HCH contact insecticide is incorporated for appropriate end uses. The 'benchmark' for performance is based on 1% w/w TBTO and 0.5% w/w organochlorine insecticide concentrations in petroleum distillate applied to give typical absorptions of 20 to 24 litres/m^3 in joinery components. At these concentrations they are not effective against soft rots and therefore can only be used in situations of low or moderate risk typically above damp-proof course level.

3 THE ECONOMICS

Wood has been an article of commerce for many centuries and in the past high transportation costs were justified only for the most valuable woods. Wood was used principally in forest areas where it was readily available; degradation was accepted and replacement was comparatively simple and inexpensive. In contrast wood is now

Table 2 Production of Preservative Treated Wood 1988/1989 M^3

	Creosote	Salt Treated	Organic Solvents	Total
Europe	1,050,000	3,800,000	1,150,000	6,000,000
Africa	10,000	175,875	5,682	191,557
North America	2,560,612	15,751,980	1,354,693 (1)	19,667,285
Latin America	93,500	660,200	-	753,700
S.E. Asia	22,525	513,780	6,990	543,295
Japan	79,000	333,000	1,341 (1)	413,341
Australia	35,000	750,000	28,000	813,000
New Zealand	6,500	820,000	18,000	844,500
	3,857,137	22,804,835	2,564,706	29,226,678

(1) Pentachlorophenol

a valuable commodity and it is essential for it to be utilised efficiently by industrial wood preservation in order to conserve world resources.

Construction timber imported into Great Britain in the 19th Century was generally slow grown and had the sapwood removed. Dwindling resources and increased demand have resulted in more severe forestry methods to encourage maximum yield, giving rapidly grown trees with wide growth rings and composed mainly of early wood.

As the trees are felled when comparatively small in diameter a large proportion of the wood is now sapwood; Swedish redwood for example is now approximately 50% sapwood. Whilst the strength properties are not significantly affected by wide rings and the presence of a high percentage of sapwood, the wood is far less durable than the slow grown trees which contained a high proportion of heartwood which were commonly used in the past.

The value of the wood preservation industry on a world-wide basis can be demonstrated by the study of available statistics (see Table 2).

Sources include Western European Institute for Wood Preservation (WEI) and national wood preservation associations for treatments during 1988.

Based on WEI European estimates the total added value of industrial treatment is 1.3×10^9 ECU/year producing goods with a total value of 6.7×10^9 ECU/year.

Table 3 Softwood Timber Component Cost Analysis

Timber Component	Estimated Average Cost	Timber Treatment Type DV or V/P	Average Treatment Cost	Treatment as a % of Total Cost	Estimated Replacement/ Repair Cost	Treatment Cost Benefit
1) Window 1200 x 1200	£100.00	Double Vacuum Organic Solvent	£1.17	1.2%	£450.00 (replace)	£448.83
2) External Panel Door	£164.00	Double Vacuum Organic Solvent	£3.37	2.0%	£350.00 (replace)	£346.63
3) Roof Truss (8m Span)	£ 26.00	Double Vacuum Organic Solvent	£2.30	8.8%	£190.00 (repair)	£187.70
		Vacuum Pressure Waterborne	£1.70	6.5%		£188.30
4) Floor Joist 50 x 150 x 5m	£ 21.00	Vacuum Pressure Waterborne	£1.12	5.0%	£150.00 (repair)	£148.88
5) Cladding Shiplap (10 m^2) 19 x 150m	£116.00	Double Vacuum Organic Solvent	£9.11	7.28%	£276.00 (replace)	£266.89
		Vacuum Pressure Waterborne	£5.66	4.65%		£270.34
6) Fence Post 8' x 75 x 100	£ 10.05	Vacuum Pressure Waterborne	£0.51	4.8%	£ 45.00 (replace)	£ 44.49

Costs based on UK market 1990

Perhaps more simplistically the value of pretreatment can be demonstrated by comparing the cost of treatment against estimated replacement or repair cost (Table 3).

4 FUTURE TRENDS

Since the birth of the modern industrial preservation industry in 1838 much practical research and development has been carried out to bring the industry to where it stands today, principally heavy oil based preservatives for utilities, CCA preservatives for constructional products, and light organic solvent based preservatives for joinery products.

Any new industrial wood preservative which is to be successful commercially over a long period must meet the four most important criteria satisfied by these traditional products.

- efficacy against fungi and insects appropriate to the hazard of the end use for which it is destined

- it must be capable of being applied in a way not to cause any detrimental effect to the user or to the environment

- the protected timber must be safe to handle both at the treatment site and by the end user

- it must be cost effective when compared with alternatives such as plastic, concrete, aluminium or steel

Because of the need to be toxic to fungi and insects it is highly likely that any new chemical wood preservative will pose problems to the environment if applied incorrectly. Whereas the pressure processes developed in the early nineteenth century have not changed; and are unlikely to radically change into the twenty-first century, on-site environmental control will continue to become more stringent. The full implementation of European environmental law through national legislation such as the UK Government's Environmental Protection Act 1990 will mean that by the year 2000 anything less than total containment will be outlawed. This will be achieved through better designed installations operated by expertly trained staff. Controlled fixation systems will be common as will solvent removal and recovery units. The capital costs of such installations will result in a rationalisation of the industry with the growth of treatment 'factories' and disappearance of the smaller units. This trend is currently evident in Europe and has already been realised in the USA.

The current most widely used preservatives have been developed over many years and when applied correctly hold no threat to the treater, the user of the treated wood, or the environment. However, the legacy of the industrial revolution in which traditional chemicals were applied in either uncontrolled situations and where on-site environmental control was minimal is considerable pollution. Historically this was acceptable, not just in wood preservation, but in industry as a whole. Consequently the use of traditional wood preservatives is under close scrutiny and in some countries the emphasis is on the introduction of alternatives rather than the improvement of application techniques and on-site control. These pressures are likely to bring about fundamental changes in wood preservation systems over the next twenty years comparable with the pace of development through the early nineteenth century and against the stability of the past twenty years.

Although the wood preservation industry is highly significant it is not of sufficient size or adequately financed to develop specific biocides with the necessary broad spectrum characteristic essential for a successful wood preservative. The industry is therefore dependent to a great extent on testing biocides developed in other market sectors such as agrochemistry. The new 'environmentally friendly' agrochemicals tend to be organism specific and are designed to biodeteriorate rapidly in ground contact, properties not necessarily attractive for an effective wood preservation system.

Usage of creosote has relatively stabilised over the past few years and this is not anticipated to change greatly in the foreseeable future. Although creosote is an excellent preservative it suffers the disadvantages of smell and exudation from the surface of the treated wood and therefore its use is likely to be restricted to treatment of specific utilities such as telegraph poles or railway sleepers. Attempts to overcome these inherent properties will continue to make creosote treated timber more user friendly and water emulsified creosote systems have been developed.[13,14]

However, the main changes in the use of creosote will focus on plant design and quality of treatment with the emphasis on total containment on site and quality assured treatments. Within Europe pressures are growing to use 'cleaner' creosote with reduced benzopyrene levels and it is possible that these moves will meet with some success. However, the effectiveness of the preservative could be adversely affected if these changes are too severe. The use of creosote at DIY level is also under threat in some countries and with the growth of environmental consumerism and regulation it is likely that this use will disappear or be severely restricted.

During the past twenty years the main growth areas have been in waterborne treatments, principally in the USA based on the increased use of CCA-treated timber in commercial and domestic DIY outlets. It is this area which has greatest potential for growth if oil prices suffer instability or environmental pressures on the use of solvents in joinery applications increase and where the potential for change in the nature of the preservatives is the greatest. Planned punitive 'environmental' levies and legislation restricting the use of traditional products based on creosote or chromium and arsenic, in Scandinavia are designed to make their use uneconomic in low hazard situations. Unless industry can develop and introduce suitable alternatives which are perceived to be more environmentally acceptable there could well be a move away from timber towards the use of alternative materials such as steel, aluminium or plastic. The ecological merit of such a move is highly questionable.

Due to the excellent track performance and inherent safety record of the traditional preservatives the industry faces a great challenge in introducing commercially viable and effective alternative products.[15]

Of the commercially available waterborne preservatives it is the arsenic pentoxide component of CCA which attracts most attention, perhaps due to the well known reputation of the toxicity of 'arsenic'. Preservatives based on copper and chromium which do not contain arsenic have been available for some time although new and different forms such as oxide liquors are still being introduced.

The most common of these products are copper-chromium (CC), copper-chromium-boron (CCB), copper-chromium-fluoride (CFK), copper-chromium-fluoroborate (CCFB) and copper-chromium-phosphorous (CCP). The additional component is included where appropriate to protect timber from degrade by copper tolerant basidiomycete fungi. Comparison of the performance of copper-chrome based preservatives against Coniophora puteana[16] indicates the effectiveness of the added boron in CCB although it is clear from the leached data that this component is not well fixed after leaching. Although CCP is well fixed the results highlight the reduction in performance compared to CCA and CCB and the ineffectiveness of the phosphorus component in preventing attack against Coniophora puteana (Table 4).

Laboratory leaching experiments have also illustrated the lack of permanence of the fluoride and boron components in CFK[10] and CCB[7] and confirmed in comparitive tests.[17]

It can be concluded that the replacement of arsenic by other components has only been a partially effective means of improving

Table 4 Toxic Threshold Values of Copper-Chrome Based Preservatives against Coniophora puteana

Product	Toxic Limits (kgm^{-3} product) Unaged	Leached
CCA	0.8-1.8	0.8-1.8
CC	16.0	-
CCB	3.1-6.3	17.9-23.9
CCP	11-18	11-16

the overall performance where long-term performance in high hazard situations is required. This does not preclude the successful use of such products in lower hazard situations where the product may be kept dry or subject only to intermittent wetting.

Increasingly, the use of chromium is coming under scrutiny and in response a second group of preservatives are entering use. These preservatives are based on copper which is likely to remain as the principal fungicidal component in most future systems. In order to bring about copper 'fixation' alternative methods have been examined which include the formation of water insoluble copper soaps such as copper naphthenate[18] or the use of ammoniacal/amino based systems as a means of depositing the active element in the wood.[19]

Ammoniacal copper formulations have a long history of use and can be effective in many situations. However their use has generally been limited because of problems with corrosivity, noxious odours and hygiene at the treatment plant, and perhaps more importantly the unsightly appearance of treated wood.

The amino copper range of formulations are similar in many ways but offer the advantages of lower odour and no unslightly surface deposit on the wood. One such product, containing copper Bis (N-cyclohexyl diazenium dioxide) has gained some popularity in West Germany. However, the product is very aggressive towards steel in treatment plants and the treated timber colour ages rapidly to an unattractive grey. The introduction of these products into Germany has been reviewed and calls for a cautionary approach in the replacement of traditional, well tried and proven products.[20] The review recognises the difficulty in judging the long-term performance of products based on laboratory tests and artifical ageing. It is claimed the new products are being introduced without the necessary field test support risking the future credibility of the industry.

Preservatives based on quaternary ammonium compounds (AAC) such as dialkyl dimethyl ammonium chlorides were introduced into commercial use in New Zealand in the late 70's following excellent indications of performance in laboratory tests. Unfortunately the performance in the field was very disappointing and these problems have been discussed suggesting failures due to misuse in practice, depletion of the active ingredient by chemical adsorption and possible biodetoxification, and possibly attack by AAC tolerant Coniophora spp.[21,22]

Other organic biocides which have attracted much attention as potential replacements for existing products include 3-iodo-2 propynyl butyl carbamate (IPBC), 2-(thiocyanomethylthio) benzothiazole (TCMTB), triazole derivatives, isothiazolones, dichlofluanids and sulphamides. Although laboratory tests have indicated good fixation and fungicidal performance long-term field performance, has not been proven, particularly in high hazard situations.

Although the properties of borates have long been recognised they are not currently used in significant quantities. However, due to their inherent properties of a wide spectrum of fungicidal and insecticidal efficacy, coupled with a relatively low mammalian toxicity there has been an increase of interest in their use. Borates have one major disadvantage in that they do not chemically fix to the wood structure and so are prone to leaching under exposed conditions. Attempts to overcome this disadvantage have so far been unsuccessful and therefore their use has been restricted to specific end uses such as certain constructional timbers or remedial treatments.[23]

Commercial diffusion treatments are widely used in Australasia with thickened systems gaining popularity in recent years. Vapour impregnation of dry timber using trimethyl borate is currently under investigation and this may be applicable in the production of board products such as medium density fibre board (MDF) and orientated strand board (OSB).[23,24] Solid borate pellets or rods have gained commercial acceptance in the remedial treatment of poles, railway cross ties and more recently in remedial joinery products.

Likely trends in the developments of LOSP active ingredients have been reviewed and cite the recent introduction of the synthetic pyrethroids cypermethrin, and in particular permethrin as acceptable alternatives to the traditional organochlorine insecticides. Although TBTO and PCP remain to be the most widely used fungicides, zinc soaps such as the versatate in Europe and IPBC in the USA are gaining commercial acceptance.[11]

With the long-term uncertainties surrounding the use of

solvents attempts to develop suitable water based formulations for joinery applications will continue based on micro-emulsion technology. The requirements for an acceptable system include physical stability under vacuum and pressure, stability to wood extractives, no depletion from solution on prolonged usage and good penetration of active ingredients into the timber. Additionally disadvantages such as grain raising, dimensional change and the need for post-treatment drying and fixation will have to be overcome for these systems to gain acceptance. As yet no emulsion based system has challenged the established products mainly due to costs and practical difficulties in industrial application.

5 CONCLUSIONS

Since the invention of pressure impregnation in 1838 the industrial wood preservation industry has developed on a world-wide scale to meet the requirements of efficiently using natural non durable timber resources.

The industry has enjoyed relative stability through the 1960's to the 1980's mainly due to the excellent performance of traditional preservative systems. With the increasing concern over the protection of the environment, legislative pressures will demand total containment of all sites by the year 2000 irrespective of the preservative being used.

These measures will ensure the continued use of traditional preservatives into the 21st Century, particularly in high hazard areas. However, for lower hazard end uses there is the potential for increased growth of alternative preservative systems.

The wider range of products and processes should lead to increased research and development concentrated on better understanding of fixation mechanisms, biodeterioration and application of more novel chemical and physical methods of protection. Many concerns have been expressed about the lack of investment in research and development and it has been proposed that the winners of the commercial battle will be those who invest most and most wisely.[25] *The future success of the industry to meet the environmental challenge of the 90's and 21st Century will depend on continuing healthy technological competitiveness within the industry. Without the innovative introduction of new products and processes the constructional industries will look increasingly towards alternative materials to timber.*

REFERENCES

1. R. Smith, Hickson's School of Wood Preservation, 1979, H.T.P

2. B. Hickson, Development of the Hickson Group, 1967, H.T.P.
3. G.D. Fahlstrom, et al US For. Prod. J, 1967, 17, No.7: 17.
4. W.J. Henry and E.B. Jeroski, Proc. A.W.P.A. 1967, 63 : 187.
5. A. Wilson, J. Inst. of Wood Sci., 1971, 36.
6. J. Dunbar, Proc. A.W.P.A. 1962.
7. G. Becher and C. Buckmann, Holz-forschung, 1966, Bd20 Heft 6.
8. M.E. Hedley, Proc. N.Z. W.P.A. 1984, 24, 48.
9. M.R. Gayles, H.T.P. Int. Report W1.2/21, 1991.
10. G. Deon, Mat. and Org., 1973, 8, 4; 296.
11. D.A. Lewis and D. Aston, Proc. B.W.P.A, 1989.
12. J.C. Beech and P.L. Newman, B.R.E. Paper, 1975, 97/75
13. C.W. Chin, J.B. Watkins and H. Greaves, IRG Doc. IRG/WP/3235, 1983.
14. H. Greaves, Ann. Rev. Div. Chem. & Wood Technol, 1986, 25
15. M. Connell, J.A. Cornfield and G.R. Williams, IRG/WP/3573, 1990, 320.
16. R.J. Tillott and C.R. Coggins, Proc. B.W.P.A. 1981, 32.
17. D.G. Anderson, Proc. C.W.P.A. 1989.
18. H.M. Barnes, For. Prod. J., 1988, 38, 10, 77.
19. M.A. Hulme, Proc. B.W.P.A., 1979 38.
20. H.N. Marx and C. Wache, Bausortiment Und Heimwerkerbedarf, 1990, No.8.
21. J.A. Butcher, IRG Doc. No. IRG/WP/3328, 1985.
22. J.N.R. Ruddick, IRG Doc. No. IRG/WP/3428, 1983.
23. D.J. Dickinson, R.J. Murphy, Proc. B.W.P.A. 1989, 35
24. P. Vinden, T. Fenton, K. Nasheri, IRG/WP/3329, 1985.
25. E.A. Hilditch, J.O.C.C.A. 1990, 10, 405.

Diffusion Treatment of Wood—An American Perspective

Lonnie H. Williams

USDA FOREST SERVICE, SOUTHERN FOREST EXPERIMENT STATION, PO BOX 2008 GMF, GULFPORT, MISSISSIPPI 39505, USA

1 INTRODUCTION

Although some of the early research and commercial use of diffusible preservatives occurred in the United States, most of the research and commercial use of diffusion treatments has taken place elsewhere. Only within the past decade have diffusion treatments received significant research and commercial attention in the United States--the world's leading market for wood consumption.[1] This increased activity offers important potential markets for diffusible chemicals and diffusion-treated wood in the United States as well as other countries, stimulating research and commercial development of diffusion treatment technology.

Many factors affect diffusion, and numerous options exist for accomplishing treatment. Thus there are many opportunities to improve the use of diffusion treatment technology to protect various wood products.

This paper provides an overview of diffusion treatment technology, reviews efforts to enhance its use, describes some advantages and limitations of diffusible preservatives, and discusses issues possibly inhibiting or aiding use of diffusion treatments in the Western Hemisphere.

2 DIFFUSIBLE PRESERVATIVES, PRINCIPLES, AND PROCESSES

Preservatives

In its simplest form, diffusion treatment is a natural process in which molecules of the preservative migrate through a porous medium, wood. Diffusion is based on the tendency of certain compounds to equalize differences in concentration of solutes. Examples of diffusible preservatives include: sodium arsenate, borax and boric acid (usually used in combination), disodium octaborate tetrahydrate, copper sulphate, sodium fluoride, potassium fluoride, and zinc chloride.

Principles

Wood can be penetrated by a water-borne preservative without any liquid absorption if the wood cells contain free water that can make contact with the external preservative (cell luminas contain free water; cell walls contain bound water).[2] Then, ions or molecules of the dissolved preservative seek an equilibrium concentration by random motion along the liquid pathways in wood in reasonable accord with Fick's laws of diffusion. The resulting penetration is independent of any solution absorption. Because wood is a nonhomogeneous material and is seldom totally saturated with water when treated, the "traditional" diffusion treatment probably involves both absorption and diffusion. In addition, the effects of chemical interactions between preservatives and with the wood cannot be ignored. Details on mathematical principles and theories related to diffusion treatment of wood or bamboo may be found in a number of references.[3-12]

Processes

Preserving wood by diffusion with aqueous preservative salt systems consists of two steps: 1) getting a sufficient quantity (loading) of the preservative on the surfaces of wood with a sufficient moisture content to allow diffusion, and 2) storing treated wood under proper conditions for sufficient time until the wood is penetrated by the preservative to the desired depth. The term "dip-diffusion" is often used to describe the treatment process whereby unseasoned wood is momentarily dipped in heated, concentrated borate solutions and then stored under restricted drying conditions for penetration by diffusion.

Diffusion treatment with water-based preservatives such as borates can be done in many ways. Borates can be applied to wood by dipping, spraying, soaking, compression rolling, and as thickened solutions before diffusion storage. Diffusion penetration can be enhanced and storage time shortened by first dipping wood in a hot solution (e.g. 80°C) and then in a solution at ambient temperature. The same effects can be obtained by soaking wood in a heated treating solution. Borates also can be applied as a vapor, and by pressure treatment of unseasoned or dry wood. In addition, borates can be incorporated into various wood-based, foam, or plastic products during the manufacturing process. Both borates and fluorides can be applied for remedial treatment as pastes, solid rods or capsules, and bandages.

Diffusion treatment also includes treatment of poles and other large timbers with volatile compounds.[13,14] Transported through wood in a gaseous phase, these compounds arrest attack by decay fungi and insects. Commonly used compounds include: Chloropicrin (trichloronitromethane), sodium n-methyldithiocarbamate, and methylisothiocyanate. These treatments can provide protection that lasts for years.

Recent efforts with fluoride preservatives have focused on the use of bi-fluorides, such as ammonium hydrogen fluoride, which produce free hydrogen fluoride in wood. This allows deep penetration of fluoride in the vapor phase. In several European countries, this technology is used as both a pretreatment and a remedial injection treatment for joinery *in situ*. Reviews of fluoride wood treatment technology provide additional information.[15,16]

Diffusion treatment with aqueous preservative systems usually results in a concentration gradient of the preservative, with higher concentrations near the wood surfaces and lower concentrations deeper within wood. This "window-frame" pattern of penetration is typical when there is little or no absorption of solution. Depending on the conditions of treatment or final use, it is theoretically possible for a uniform, equilibrium concentration to be reached in wood.[17] This result, however, does not usually occur in practice.

Pressure treatment of dry wood with nondiffusible preservatives also usually produces a concentration gradient of preservative within wood. This is particularly true for refractory (difficult-to-treat) wood such as the heartwood portions of spruce (Picea spp.) or Douglas-fir (Psuedotsuga menziesii [Mirb.] Franco). However, the general public and the construction industry often mistakenly assume that pressure-treated wood is uniformly and thoroughly penetrated with preservative. It is surprisingly difficult to convince the uninformed that refractory wood can be more completely treated while unseasoned by simply dipping it in an aqueous diffusible preservative solution and storing it for a sufficient period.

Commercial kiln drying can alter the penetration patterns in diffusion-treated wood, and subsequent machining operations can expose wood centers with insufficient preservative retentions. Therefore, treatment specifications should prescribe the minimum retention of preservative needed at a given depth in wood, according to the final use of the wood, for protection against fungal and insect damage. Factors affecting treatment also should be given proper consideration and be controlled as much as possible.

3 FACTORS AFFECTING DIFFUSION

Factors affecting diffusion by aqueous preservative systems include: temperature during diffusion storage, length and method of storage, solution concentration, and type of preservative. Other factors include characteristics associated with the wood being treated such as moisture content, grain direction, density, sapwood to heartwood ratio, and surface characteristics.[2,18] Of these, the most important factors affecting the rate of diffusion probably are temperature, wood moisture content, wood density, and type of preservative. The remaining factors tend to affect the uniformity of treatment.

Temperature

The rate of diffusion increases as temperature increases and can be expected to nearly double with each 20°C rise in temperature during diffusion storage, thus halving the length of the storage needed to obtain the same results.[19,20] For radiata pine (*Pinus radiata* D. Don) stored at 60°C instead of 20°C, there is a four-fold increase in the rate of diffusion, reducing the diffusion storage time by three-quarters.[21]

The effect of temperature is one reason diffusion treatment with borates is suitable for tropical hardwoods. Diffusion penetration of low density hardwoods can occur quite rapidly in the constant high temperature and humidity conditions of the tropics. For example, 38 mm-thick banak (*Virola* spp.) wood in the Brazilian Amazon is completely and adequately penetrated after 7 days of storage in a shed following a 1-minute dip in a 25 percent boric acid equivalent (BAE) solution.[22]

Wood Moisture Content

The moisture content of the wood being treated is a critical factor because diffusion with aqueous-based preservatives begins at about the fiber saturation point (about 30 percent wood moisture content). In practical terms, diffusion is best when moisture content is much higher. The rate of diffusion increases rapidly with increases in wood moisture content up to about 100 percent, but increases above this level have little effect.[23] Because wood moisture content is based on the oven-dry weight of wood, levels as high as 240 percent are possible with some species. An understanding of the effect of moisture content helps explain why penetration results are different for sapwood and heartwood and when treating wood already in use. Heartwood and wood in service usually have much lower moisture contents.

Wood Density

An inverse relationship exists between the density of wood and the rate of diffusion, but few studies of this relationship have been reported.[2] When comparing treatment of woods with air dry specific gravities of 0.71 and 0.96, the transverse diffusion rates were about 4 to 6 times faster in the lighter wood and almost 17 times faster in the longitudinal direction.[4] The great effects of density on diffusion are especially important when treating a mixture of species with varied densities.

Type of Preservative

Despite the 50-year history of diffusion treatment technology, considerable opportunities exist to enhance preservative formulations specifically for diffusion treatment or to retard leaching. Perhaps this could be accomplished by studying the influences and reactions between chemicals within the wood substrate and the preservative.[2,16] Interestingly, there appears to be a 39-year gap between the published papers about the reactivity between the substrate and diffusible preservatives.[4,24]

In 1951, Christensen found that the rate of diffusion through wood of solutes containing either a bivalent or trivalent anion or cation can be 3 to 10 times slower than that of electrolytes in which both ions are univalent. He showed that the restricted rate appears to be greatest in the heartwood of dense timbers, where pathways often are narrow or partly blocked by chemical deposits. The anion was the restricted ion. This may explain why solutes not disassociated often diffuse rapidly--because they move through wood as electrically neutral molecules.[2]

In 1990, Lloyd and coworkers found that a borate preservative chemically reacts with poly-ols of wood, thus affecting its toxicity to certain organisms.[24] This discovery raises the possibility that the rate of diffusion for borates could possibly be aided by formulation enhancement of chemical reactions. This also suggests that the rate of leaching could be reduced in certain wood species by specific borate formulations.

These discussions of factors affecting diffusion are generalizations based on test results with different wood species and preservatives. Therefore, it may be useful to reexamine influences affecting treatment of a specific wood resource with a specific water-based preservative.

4 WORLDWIDE USE OF DIFFUSIBLE PRESERVATIVES

Excellent reviews have recently been made of the historical development and use of diffusible preservatives in Australasia, Europe, and North America.[16,25,26] Canadian experiences with diffusion treatments also have been described.[27]

Diffusion treatments to protect building timbers have been developed to their greatest potential through research and commercial development which first began in the late 1930's and 1940's in Australia and New Zealand, respectively. In Australia, diffusion treatments first were done to protect hardwoods from attack by destructive powderpost beetles (Coleoptera: Lyctidae). A complex boron-fluoride-chromium-arsenic (BFCA) formulation with a high solubility in cold water was developed. In 1964, this formulation was approved for treatment of all sawn building timbers in Papua New Guinea, where it is still in use today.[2,28] Its use in New Guinea, including health hazards and field performance, has been extensively described.[29]

In New Zealand, commercial diffusion treatments were done to protect softwood building timbers. These treatments were largely based on research conducted by Harrow, who extensively studied such variables as diffusion time, solution concentration, temperature, and timber thickness.[6,7,9,30,31] Recently, various aspects of diffusion treatment were reexamined as well as criteria affecting the economics of the process.[12]

In North America, early work focused on dip-diffusion treatment of poles and mining timbers, double diffusion treatment of posts, and remedial groundline treatment of poles.[26] Today, commercial use strongly continues for the latter. Farmers once widely used double diffusion treatments, first treating posts with copper, zinc, or nickel sulfate and then with sodium dichromate, sodium arsenate, or both. These treatments were supplanted by the commercial pressure-treating of posts with creosote, pentachlorophenol, or chromated copper arsenate (CCA). Use of all these chemicals by individuals has now been banned in North America.

In Europe, development of the use of fluorides began in Germany in the early 1900's and now continues with the use of bifluorides, particularly in the Netherlands.[16] Efforts were made in the United Kingdom during the 1960's to market boron diffusion-treated building timbers. Because most of the timbers were imported, treatment had to be done by timber producing countries such as Finland and Canada. Problems with supply and stocking of treated timber resulted and the marketing effort failed. The use of remedial treatments with borates and other diffusible preservatives continues to grow.[32]

5 METHODS OF ACCELERATING DIFFUSION

Many efforts have been made to enhance treatment with borates during the 1960's and 1970's in New Zealand without much commercial success. For example, pressure treatment of unseasoned wood was tried by first steaming the wood and using a vacuum to reduce its moisture content before using alternating pressure cycles during treatment.[21,33,34] Other attempts involved steaming wood and immersing it in a cold solution followed by diffusion, or heating wood in diffusion storage.[35,36] These efforts led to two significant enhancements in the preservative system.

The first was the addition of a thickening agent--polymer--to the borate solution, as well as a fungicide to control the growth of mold and mildew fungi during diffusion storage.[12,37] Reported advantages include: More uniform uptake and distribution of the preservative within timber, faster diffusion times, lower energy costs (no heating of solutions required), improved control of fungi during diffusion storage (fungicide does not breakdown because it's not heated), and better treatment of surface-dry timber. About half of the treating plants in New Zealand switched to using the thickened borate product within its first year of commercial availability. The impact of this technology in other parts of the world remains to be determined.

The second enhancement combines the steps of borate treatment and drying into one operation.[38] The process is based on the fact that organoboron compounds readily hydrolize with the moisture present in wood to form boric acid. First, solid wood is high-temperature dried to about 3 percent moisture content. Next, trimethyl borate is introduced into the hot kiln/cylinder where it vaporizes and penetrates the wood as a gas. Then, the wood is steam reconditioned back to the desired moisture content while under heavy weights. The entire process--from unseasoned timber to a dry treated product--takes about 30 hours.

This technology appears particularly suitable for treatment of wood-based composite products such as oriented-strand board, plywood, and waferboard which have a low moisture content at various times during their manufacture. This application is being tested now at London's Imperial College and the New Zealand Forest Research Institute.[39]

6 ADVANTAGES AND LIMITATIONS OF DIFFUSIBLE PRESERVATIVES

Diffusible preservatives, especially the borates, have some outstanding characteristics for protecting wood in buildings. However, use of the fluorides may be limited in the United States to certain specialty applications and delivery systems because of their toxicity, volatility, and corrosiveness. The beneficial characteristics of borates include: toxicity toward most wood-damaging fungi and insects; mobility to reach decayed- and subterranean termite-damaged areas; low toxicity to mammals; and adaptability to a wide range of treatment processes.[40] And borates reportedly cause no odor, color, corrosiveness, or wood machining problems, are cost effective, and increase the resistance of wood to fire. In addition, borates allow good treatment of large timbers and refractory woods while they are still unseasoned.

Many of the described advantages of borates accrue when treating unseasoned wood. This restriction, as well as long diffusion times, often hinders the use of borates. The producer should be treating unseasoned wood, but producers often do not have the same incentive to treat as those wanting to use a treated product. Leaching is sometimes regarded as a major disadvantage to using borate-treated wood. Dip-diffusion treatments often are regarded as being too simple and uncontrollable. Problems may result because additional fungicides are required when treating unseasoned wood.

Thus, diffusion treatments with borates present several paradoxes. The simplicity of borate treatment procedures and equipment is highly advantageous for use in tropical climates and developing countries. But simple equipment and procedures are not suitable for high volume production and the production-on-demand markets of developed countries. The mobility of borates in wood is highly advantageous when treating large timbers, refractory woods, and wood in service. But the mobility and leachability of borates results in unsuitable protection for wood in ground contact or frequently wetted, unless additional protective coatings or preservatives are used. Borate treatment makes wood more fire resistant, which is also advantageous when treating building timbers in service. Whether the low levels of borates needed for insect and fungal protection provide significant fire protection in buildings constructed entirely from borate-treated wood remains to be determined. However, the added fire resistance can create a disposal problem when waste is burned for power generation.

7 STATUS OF DIFFUSION TREATMENT USE--WESTERN HEMISPHERE

Borate wood preservatives are not currently in wide use in the Western Hemisphere. For the United States, suggested reasons for this lack of use were: (1) early U.S. test results "branded" borates as ineffective preservatives because they did not protect wood in ground contact, which is the major use of preservative-treated wood then and now; (2) U.S. structures were being protected from the major insect threat--subterranean termites--with chlorinated insecticide soil treatments before borates were shown in some foreign countries to be toxic to termites and effective for protecting building timbers; and (3) a mobile U.S. society with abundant wood resources placed low priority on long-term prevention of damage by decay fungi and insects to buildings.[41] However, prevention through construction with treated wood has become a common practice in the U.S. Virgin Islands and Hawaii where insect and decay hazards are great.

In the 1960's, diffusion treatment of Canadian softwoods were studied in order to supply borate-treated building timbers to United Kingdom markets, but these markets never materialized.[27] Diffusion treatments also received limited testing during the 1960's in several Latin American countries. The most important suggested reason for the limited attention to diffusion treatments in Central and South America is that wood-frame housing is held in low regard while masonry housing, with interior woodwork of naturally durable woods, is preferred.[42] Thus, no strong incentive exists to treat less durable secondary species for use in local economies.

Interest in diffusion treatments languished until it was rekindled by the results of research begun in the late 1970's at the USDA Forest Service's Southern Forest Experiment Station Laboratory at Gulfport, Mississippi. The original goal of this research was to find a preventive measure for lyctid powderpost beetle infestations in tropical hardwood materials imported from Brazil.[22,43-46] Soon, this research was greatly expanded through cooperation with researchers at the Mississippi State University Forest Products Laboratory (MFPL) and Oregon State University Forest Research Laboratory (FRL).[1]

Within the past 6 years considerable information about borates has been disseminated by the USDA Forest Service in cooperation with MFPL and FRL personnel. This dissemination effort--with the theme "Borates For Wood Protection"--has included the production and distribution of two videotapes about borates.[47] Also, the First International Conference on Wood Protection with Diffusible Preservatives was organized and held November 28-30, 1990, in Nashville, Tennessee. The proceedings of this conference revealed that many U.S. and Canadian researchers with various institutions and commercial companies currently are studying the use of diffusible preservatives.

Fourteen small borate-treated test structures have been built at Gulfport and at Starkville, Mississippi, and more full-size structures will be built at Gulfport and other governmentally supervised sites in various southern states, the U.S. Virgin Islands, and Puerto Rico. The purpose of these structures is to permit a controlled, field evaluation of the prevention of termite damage by construction using borate-treated components.

As a result of the above research and information dissemination, borates now are used commercially for treatment of: unseasoned lumber (*Virola* spp.) in Brazil, Ecuador, and Peru; logs for log-kit homes produced by several of the largest manufacturers; pallets produced for use by the food industry; some timbers and dimension stock used in housing construction in the Virgin Islands and some other locations; and some specialty products. Two large corporations that produce housing kits from stressed-skin panels (a layer of polystyrene foam between sheets of waferboard or oriented-strand board) are now treating these panels during production.

Borate diffusion treatments are now recommended for prevention of some insect problems in hardwood lumber.[48] Recommendations for using diffusion-treated building materials are included in the draft revision of the "Guidelines for Protection of Wood Against Decay and Termite Attack" produced by the Wood Protection Council of the National Institute of Building Sciences. Borates also are approved as a wood preservative by the American Wood Preservers' Association, but a treatment specification currently is available only for southern pine timbers.

8 FUTURE FOR DIFFUSION TREATMENTS--WESTERN HEMISPHERE

In America, the future for widespread use of diffusible preservatives and diffusion treatments remains uncertain, despite recent renewed interest. It seems clear that diffusible preservatives such as the borates will be much more widely used because of their favorable properties for treating a wide variety of wood products and for reducing potential hazards to the environment and users. However, the future for widespread use of traditional diffusion treatments seems very uncertain. Diffusion treatments are not likely to be used by large U.S. hardwood or softwood lumber producers until the technology is developed to avoid long diffusion times and assure a high degree of sophistication and process control over treatments.[16] Large-scale use of pressure treatment with borates by Canadian or U.S. building timber producers is also unlikely because such treatment still involves the costs of drying lumber before and after treatment.[49] Many issues about the future use of these preservatives were discussed at the recent Diffusible Preservatives Conference.[26,50,51]

Issues Inhibiting Use

A well established practice in the United States is to use products that are pressure-treated with preservatives which become "fixed" in wood for ground contact or exterior exposure. To avoid the potential hazards from using traditional wood preservatives, some borate-treated wood already has been misused-- as treated wood traditionally has been used--in ground contact. Because the borates remain subject to leaching, borate-treated wood should not be used in ground contact, except when combined with preservatives approved for that purpose. Therefore, significant technical literature and training on product use must be provided to avoid damaging the credibility of borates or any other diffusible preservatives.

Borates and diffusion-treated wood are well suited for building timbers, but industry and the public have not perceived an overwhelming need to provide long lasting preventive treatments to building timbers. Consumers do not understand the underlying causes for decay and insect problems in buildings, and are reluctant to spend construction money on treated wood to prevent problems that may occur many years later. By this time, ownership of the structure may have changed several times.

Persuasive economic, technical, and psychological reasons would help overcome the reluctance to spend money "up-front" on prevention through construction with treated wood. However, the national economic impacts of most fungal and insect problems are poorly documented.[50] For example, estimates for the most well-known insects, termites, varied nearly five-fold from $753.4 million to $3.4 billion.[52,53] National loss estimates for decay fungi and other wood-infesting insects are even less certain, and the incidence of various problems by locality or number of structures is not known.

Construction with treated wood must compete with two well established practices--insecticide treatment of soil before building construction for prevention of subterranean termites, and use of wood for construction or interior woodwork without treatment. An extensive education and marketing effort will be required to convince consumers and building protection authorities that construction with borate-treated wood is effective and more economical than the established practices.[50,51] Also, the use of soil treatments is supported by: large commercial pest control industry; federal, state, and local building codes; several large chemical companies extensively supporting their products with good distribution, marketing, technical services, and significant research funds.

Diffusible chemicals have considerable potential for providing effective preventive or remedial control of both decay fungi and insects.[16,32,54] The properties of borates--environmental friendliness, low volatility, and low mammalian toxicity--are attracting much attention from the U.S. commercial pest control industry. A marketing effort is underway to treat building frameworks at construction sites with borate alone or borate/glycol solutions, instead of purchasing borate-treated wood for construction. The final outcome of this effort is not yet known. If these treatments to dry wood give good protection, then consumers may regard building with borate-treated wood favorably. If these treatments fail, then the consumer may conclude that borate-treated wood used in construction also would be ineffective. This would be unfortunate because wood producers supplying borate-treated building materials offer the greatest benefits for wood and energy conservation while providing protection of buildings.

Issues Aiding Use

Issues that may help evoke a change toward construction with borate-treated wood were discussed:[50] increased consumer environmental awareness, problems with structural pest control practices, changes in building construction practices, and a changed wood industry environment.

Higher priority is being given to forest resource benefits other than production of wood products--benefits such as wildlife, recreation, and helping to cleanse the environment. This may restrict the production of forest products and increase their value. As a result improved protection and conservation of wood products may be viewed as more desirable.

When providing improved wood protection, environmentally compatible methods should be used, with consideration given to criteria such as: Energy consumption, long-term cost-effectiveness, and efficiency in wood conservation, as well as environmental and public safety.[50] For example, if wood products are used without treatment and later fail, causing costly remedial treatment or replacement, use of untreated wood would be less environmentally compatible than using wood products that were properly treated with environmentally friendly preservatives. It is further suggested that environmental compatibility should be considered when funding building protection research. Priority funding is being given to alternative approaches to crop production and controlling agricultural pests.[55]

Planting more trees or not cutting trees are often suggested as ways to reduce atmospheric carbon dioxide levels, but protecting wood products already in use also has a role in this issue. The volume of wood in use represents a tremendous reservoir of carbon (similar to the wood in trees but without oxygen release).[50] And this carbon reservoir is continually being increased by new construction.

Properly applied preservatives have been estimated to increase the service life of wood at least seven times.[56] Compared to using untreated wood, this added service life has resulted in an annual savings of $7.5 billion in the United States. In addition, an annual savings of $43 million has resulted from growth of trees that would have otherwise been harvested. These figures do not include the savings that might result because remedial treatments and damaged wood replacement were no longer necessary in buildings constructed with treated wood. Nor do these estimates consider the nonmeasurable savings that reduced harvesting or longer product service life might have on global climate change.

Current problems with insect control in existing structures also may favor a change toward construction with treated wood.[50] The identified problems included: (1) controlling termites with new termiticides, (2) reducing concerns about the use of chemicals around structures, and (3) gaining economical remedial control of beetle and drywood termite infestations.

All new soil treatment termiticides must pass a 2-year screening program at the USDA Forest Service Laboratory in

Gulfport, Mississippi, then remain effective in at least three field locations for 5 years for Environmental Protection Agency (EPA) registration. The currently registered termiticides have been estimated to offer no more than 10 years of totally effective protection.[57,58] Energy-intensive, "nonchemical" measures now are being commercially used for insect control in structures including high temperatures from propane heaters, freezing temperatures from liquid nitrogen, high frequency electric heat, and placing a layer of sand or basaltic rock particles of specific size beneath and around building foundations to deter penetration by termites.[50] Unlike construction with treated wood, the first three measures do not protect against reinfestation, and the last is directed only toward prevention of subterranean termites.

More energy conservation features in structures was also considered as a reason for changing to construction with treated wood.[50] These features make it more difficult to apply wood protection chemicals at the building site and contribute to wood decay problems.[59] Example of these features include additional insulation around foundations or in walls, double walls, and air exchange between living spaces and solariums.

The use of modular or factory-built housing will continue to increase because of rising construction costs and the need for affordable housing. Protection from decay fungi and insects could be better provided by construction using treated materials because the use of treated materials could be specified and supervised in a factory-controlled setting.[50] Also, borates appear to be suitable for treating composite materials, particularly by the vapor boron process,[49] and the use of waferboard, oriented-strand board, other composite boards, and laminated beams will continue to increase. Treating these materials at production centers would avoid the problems associated with applying treatments at building sites. In both cases, using treated materials could provide a market-oriented approach to resolving pest problems--an approach which was recommended in a recent public policy study.[60]

Concerns about the environment have caused forest products harvesting to be held in low regard. It was postulated that continued harvesting of forest resources might be considered environmentally compatible when products from these resources were effectively protected to serve their intended purpose for the longest possible time.[50] It is believed that such protection could be a value-added feature that will help U.S. wood products compete globally and against alternative materials. Because markets in Europe and the Far East are familiar with the properties of borate-treated wood, the export of treated lumber or housing should be received favorably. Long-term protection also should encourage continued use of renewable wood resources rather than nonrenewable, energy-intensive, substitute materials such as concrete, plastic, or steel.

Diffusion treatment technology is ideally suited for protection of secondary tropical hardwood species.[42,61] Several Central American countries have shown growing interest by recently conducting demonstrations of boron dip-diffusion treatments. A multi-year program plan is being drafted to aid economic development in Central America. Government/industry cooperative training will be given to improve forest products utilization (including wood preservation) and aid economic development in rural areas, thus leading to better tropical forestry management. If this program is funded and succeeds in getting rural populations interested in agroforestry and forest products utilization, it may serve as a model for improving forestry management in other tropical regions of the Western Hemisphere.

9 RESEARCH NEEDS AND CONCLUDING REMARKS

Research Needs

Well-known wood preservation specialists have concluded that diffusible borates have considerable potential as wood preservatives, but a significant amount of research will be required before the full potential of borates can be realized.[51] Identified research needs include: (1) reducing leachability (considered of paramount importance), (2) controlling mold growth on treated wood, (3) determining effectiveness against soft-rot fungi and termites, and (4) developing safe methods of disposal of treated wood residues.

All these research needs would not have to be resolved for use of treated wood in building construction. Therefore, the author's list of most needed research is: (1) proving the effectiveness of borate-treated wood in construction for subterranean termite prevention and control; (2) evaluating leaching from borate-treated wood by testing methods that determine realistic leaching rates rather than fixation of chemicals; (3) enhancing diffusion treatment formulations and methods in order to achieve shorter treatment times; (4) evaluating the potential environmental hazards from treatment operations, treated wood, and treated waste disposal; and (5) enhancing the use of diffusion treatments for protection of secondary Latin American hardwood species used for interior woodwork and furnishings.

Concluding Remarks

Diffusion treatment technology, which is essentially totally new to U.S. consumers, is most suitable for treating unseasoned wood. This is diametrically opposed to the longstanding knowledge and practice of U.S. wood users, who have learned through the years that treatment by pressure or remedial control chemicals penetrates better when applied to dry wood. Therefore, much research and consumer education will be necessary before diffusion treatment technology will be widely used in the United States. Diffusion treatments must show superior effectiveness or a competitive economic advantage over other measures. For example, it is likely that commonly used U.S. building timbers--impermeable species such as Douglas-fir, spruce, and true firs (Abies spp)--may be more thoroughly penetrated by diffusion treatment when unseasoned than by pressure treatment when dry.

REFERENCES

1. H. M. Barnes, T. L. Amburgey, L. H. Williams, and J. J. Morrell, 20th Annual Meeting, International Research Group on Wood Preservation, Working Group III, 1989, Document No. IRG/WP/3542, 16 pp.
2. N. Tamblyn, Treatment of wood by diffusion, In: 'Preservation of Timber in the Tropics', Ed. W. P. K. Findlay, Martinus Nijhoff/Dr. W. Junk, Whitchurch, England. Chapter 6, 1985, p. 121.
3. B. S. Bains and S. Kumar, J. Timber Dev. Assoc. India, 1979, 25(4), 34.
4. G. N. Christensen, Austral. J. Appl. Sci., 1951, 2(4), 440.
5. G. N. Christensen and E. J. Williams, Austral J. Appl. Sci., 1951, 2(4), 411.
6. K. M. Harrow, N. Z. J. Sci. Tech. Sect. B, 1951, 32(4), 28.
7. K. M. Harrow, N. Z. J. Sci. Tech. Sect. B, 1951, 32(4), 32.
8. K. M. Harrow, N. Z. J. Sci. Tech. Sect. B, 1952, 33, 471.
9. K. M. Harrow, N. Z. J. Sci. Tech. Sect. B, 1954, 36(1), 56.
10. S. Kumar and V. K. Jain, Holzforschung und Holzverwertung, 1973, 25, 21.
11. A. J. Stamm, Pulp and Pap. Mag. of Canada, 1953, 54(12), 54.
12. P. Vinden, J. Inst. Wood Sci., 1984, 10(1), 31.
13. J. J. Morrell, 20th Annual Meeting, International Research Group on Wood Preservation, Working Group III, 1989, Document No. IRG/WP/3525, 17 pp.

14. J. N. R. Ruddick, 14th Annual Meeting, International Research Group on Wood Preservation, Working Group III, 1983, Document No. IRG/WP/3253, 6 pp.
15. G. Becker, J. Inst. Wood Sci., 1973, 6(2), 51.
16. R. J. Murphy, In: 'First International Conference on Wood Protection with Diffusible Preservatives', Ed. M. Hamel, For. Prod. Res. Soc., Madison, WI, 1990, Proceedings 47355, p. 9.
17. L. H. Williams and M. E. Mitchoff, In: 'First International Conference on Wood Protection Diffusible Preservatives', Ed M. Hamel, For. Prod. Res. Soc., Madison, WI, 1990, Proceedings 47355, p. 136.
18. G. Becker, J. Inst. Wood Sci., 1976, 7(4), 30.
19. G. N. Christensen, Austral. J. Appl. Sci., 1951b, 2(4), 440.
20. B. R. Warren, D. C. Low, and R. V. Mirams, N. Z. J. Sci., 1968, 11(2), 219.
21. P. Vinden, T. Fenton, and K. Nasheri, 16th Annual Meeting, International Research Group on Wood Preservation, Working Group III, 1985, Document No. IRG/WP/3329, 20 pp.
22. L. H. Williams and T. L. Amburgey, For. Prod. J., 1987, 37(2), 10.
23. D. N. Smith and A. I. Williams, J. Inst. Wood Sci., 1969, 4(4), 3.
24. J. D. Lloyd, D. J. Dickinson, and R. J. Murphy, 21st Annual Meeting, International Research Group on Wood Preservation, Working Group Ia, 1990, IRG/WP/1450, 21 pp.
25. H. Greaves, In: 'First International Conference on Wood Protection with Diffusible Preservatives', Ed. M. Hamel, For. Prod. Res. Soc., Madison, WI, 1990, Proceedings 47355, p. 14.
26. W. S. McNamara, In: 'First International Conference on Wood Protection with Diffusible Preservatives', Ed. M. Hamel, For. Prod. Res. Soc., Madison, WI, 1990, Proceedings 47355, p. 19.
27. J. N. R. Ruddick, In: 'First International Conference on Wood Protection with Diffusible Preservatives', Ed. M. Hamel, For. Prod. Res. Soc., Madison, WI, 1990, Proceedings 47355, p. 58.
28. N. Tamblyn, S. J. Colwell, and G. N. Vickers, In: 'Proceedings 9th British Commonwealth Forestry Conference', 1968, 10 pp.
29. C. R. Levy, S. J. Colwell, and K. A. Garbutt, 4th Annual Meeting, International Research Group on Wood Preservation, 1972, Document No. IRG/WP/310, 67 pp.
30. K. M. Harrow, N. Z. Timber J., 1955, 2(4), 28.
31. K. M. Harrow, N. Z. Timber J., 1955, 2(4), 38.

32. D.J. Dickinson, In: 'First International Conference on Wood Protection with Diffusible Preservatives', Ed. M. Hamel, For. Prod. Res. Soc., Madison, WI, 1990, Proceedings 47355, p. 87.
33. P. Vinden, 18th Annual Meeting, International Research Group on Wood Preservation, Working Group III, 1987, Document No. IRG/WP/3439, 10 pp.
34. M. J. Collins and P. Vinden, 18th Annual Meeting, International Research Group on Wood Preservation, Working Group III, 1987, Document No. IRG/WP/3438, 9 pp.
35. A. J. McQuire, N. Z. Timber J., 1961, 7(12), 63.
36. A. J. McQuire and K. A. Goudie, N. Z. J. For. Sci., 1972, 2(2), 165.
37. P. Vinden, J. Drysdale, and M. Spence, 21st Annual Meeting, International Research Group on Wood Preservation, Working Group III, 1990, Document No. IRG/WP/3632, 8 pp.
38. R. Burton, T. Bergevoet, K. Nasheri, P. Vinden, and D. Page, 21st Annual Meeting, International Research Group on Wood Preservation, Working Group III, 1990, Document No. IRG/WP/3631, 7 pp.
39. P. Turner, R. J. Murphy and D. J. Dickinson, 21st Annual Meeting, International Research Group on Wood Preservation, Working Group III, 1990, Document No. IRG/WP/3616, 13 pp.
40. D. M. Carr, In: 'Proceedings 1959 Convention of the British Wood Preserver's Association', 1959, 18 pp.
41 L. H. Williams, T. L. Amburgey, and B. R. Parresol, In: 'First International Conference on Wood Protection with Diffusible Preservatives', Ed. M. Hamel, For. Prod. Res. Soc., Madison, WI, 1990, Proceedings 47355, p. 129.
42. L. H. Williams, In: 'First International Conference on Wood Protection with Diffusible Preservatives', Ed. M. Hamel, For. Prod. Res. Soc., Madison, WI, 1990, Proceedings 47355, p. 29.
43. L. H. Williams, Integrated protection against lyctid beetle infestations. Part I. The basis for developing beetle preventive measures for use by hardwood industries. USDA Forest Service, Southern Forest Expermiment Station, New Orleans, LA, 1985, 12 pp.
44. L. H. Williams and J. K. Mauldin, For. Prod. J., 1986, 36(11/12), 24.
45. H. M. Barnes and L. H. Williams, For. Prod. J. 1988, 38(9), 13.
46. H. M. Barnes, and L. H. Williams, For. Prod. J. 1988, 38(9), 20.

47. W. H. Sites, In: 'First International Conference on Wood Protection with Diffusible Preservatives', Ed. M. Hamel, For. Prod. Res. Soc., Madison, WI, 1990, Proceedings 47355, p. 95.
48. H. B. Moore, J. D. Solomon, S. J. Hanover, and M. P. Levi, 'Lumber defects caused by insects, fungi, and chemical stains: A guide for native southeastern woods used in furniture manufacturing', Agricultural Extension Service, North Carolina State University, Raliegh, NC, 1990, Publication AG-425, 57 pp.
49. P. Vinden, In: 'First International Conference on Wood Protection with Diffusible Preservatives', Ed. M. Hamel, For. Prod. Res. Soc., Madison, WI, 1990, Proceedings 47355, p. 22.
50. L. H. Williams, In: 'First International Conference on Wood Protection with Diffusible Preservatives', Ed. M. Hamel, For. Prod. Res. Soc., Madison, WI, 1990, Proceedings 47355, 43.
51. D. D. Nicholas, L. Jin, and A. F. Preston, In: 'First International Conference on Wood Protection with Diffusible Preservatives', Ed. M. Hamel, For. Prod. Res. Soc., Madison, WI, 1990, Proceedings 47355, p. 121.
52. J. K. Mauldin, 'Economic Impact and Control of Social Insects', S. B. Vinson, Praeger Publishers, New York, 1986, Chapter 4, p. 130.
53. U.S. Dept. of Agriculture, Insects affecting man and his possessions: Research needs in the southern region, Joint Task Force Report of the South, Region Agricultural Experiment Stations and U.S. Dept. of Agriculture, 1974, 34 pp.
54. E. L. Schmidt, In: 'First International Conference on Wood Protection with Diffusible Preservatives', Ed. M. Hamel, For. Prod. Res. Soc., Madison, WI, 1990, Proceedings 47355, p. 91.
55. B. Hileman, Chem. Eng. News, 1990, 25.
56. D. D. Nicholas and R. Cockcroft, 'Wood preservation in the USA', STU Information no. 288, National Swedish Board for Tech. Devel., 1982, 130 pp.
57. C. D. Mampe, Pest Cont., 1990, 58(2), 42.
58. J.K. Mauldin, S. C. Jones, and R. H. Beal, 18th Annual Meeting, International Research Group on Wood Preservation, Working Group 1b, 1987, Document No. IRG/WP/1323, 20 pp.
59. R. D. Graham, For. Prod. J., 1980, 30(2), 17.
60. L. Ember, Chem. Eng. News, 1989, (1), 22.
61 C. R. Levy, 13th Annual Meeting, International Research Group on Wood Preservation, Working Group III, 1982, Document No. IRG/WP/3177, 10 pp.

In-situ and Supplementary Treatments Using Solid Water Diffusible Preservatives

W. Beauford

BIO-KIL CHEMICALS LIMITED, BRICKYARD INDUSTRIAL ESTATE, GILLINGHAM, DORSET SP8 4BR, UK

1 INTRODUCTION

The improved serviceability of wood has been the philosophy behind wood preservation for nearly four centuries. In this chapter the present use of in-situ and supplementary preservation treatments, and particularly solid water diffusible chemicals, will be discussed with regard to the role they have to play in the current international wood treatment industry together with the benefits which are dervied from their application.

In the past a variety of terms, such as in-situ, supplementary and remedial have been used to describe wood preservation activities beyond the industrial or primary pressure treatment processes. The term secondary treatments will now be used to embrace these terms.

The Need for Secondary Treatment

The need for secondary treatments may appear unnecessary to many observers when it is realised that there are well established primary pressure treatment processes which have been used, in some cases, for over 150 years. The long term effectiveness of primary processes is reduced, however, by the limited penetration of chemicals into the heartwood which is difficult to treat. For instance creosote, an excellent wood preservative, does not reach the heartwood of engineering timber commodities such as railway sleepers and utility poles. This leaves the heartwood at risk from fungal attack at any time and can lead to the premature failure of the commodity in service. Because of this risk and subsequent failure, there are regular inspections and replacements of wood sleepers and poles in the railway and transmission industries to maintain safe and stable infrastructures.

It has been recognised, therefore, that there is a need to provide a secondary treatment to extend the service life of wooden commodities. In the utility pole industry various methods of injecting pastes and water diffusible chemicals, such as fluorides, have been used. In recent years, the use of inorganic borates have become more appropriate and widely used because of a number of valuable properties of the preservatives:-

* Diffusibility
* Wide spectrum of biological activity
* Ability to formulate in a variety of forms

These important properties have made their application suitable for a wide range of environments and commodities. In addition, the low mammalian toxicity and general environmental acceptability of the inorganic borates are immensely valuable under current health and safety legislation where chemicals are used in-situ without the defined control of primary processes.

With any treatment philosophy it has been important to understand the reason for reduced performance. An understanding that fungal decay occurs in heartwood (which is initiated by suitable moisture conditions), the water relationships in the wood and associated technology have been as important as the development of the chemical preservation products. This is particularly important in secondary treatments which tend to use selected amounts of chemicals which are targetted to the high risk or decay susceptible areas of a wooden structure and where a number of application techniques can be employed.

Water Relationships in Wood

Wood differs from most materials used in civil engineering and construction in that it is hygroscopic and will continually exchange moisture with its surroundings,[1]. This applies in both the living tree as well as under conditions of service. The extent of the water relationships in wood are described elsewhere,[2].

In terms of both chemical preservation and its service performance the available moisture content in wood is the most important. Microbiological degradation can only occur if the wood has a moisture content exceeding 20-22% of its ovendry weight. Decay fungi can cause serious damage only when the moisture content is above the fibre saturation point (approximately 28-30%). In broad terms water can be found in wood in three basic forms:-

* Bound water in the cell walls
* Free or capillary water in the cell cavities
* Water vapour in the cell cavities

The fibre saturation point is defined as the moisture content at which cell cavities are empty of liquid water but cell walls are still saturated with bound water.

Free or capillary water is usually removed through seasoning to below the fibre saturation point, to provide material suitable for the primary preservative treatments or indeed direct use in construction and civil engineering. However, water can re-enter the wood through various pathways, particularly in external situations, to recharge the cell cavities to levels at or above the fibre saturation point. In this condition any untreated wood, such as the heartwood, is at risk from decay fungi. Decay fungi utilize diffusible enzyme systems in the free water in the cell cavities to catalyse the dissolution of cell wall substances, such as cellulose. This causes progressive changes in the structural properties of wood which will eventually lead to loss of performance.

Water Movements in Wood

It has already been mentioned that water in wood is rarely in static equilibrium as there are continual adjustments to changes in its environment. An understanding of this dynamic movement of water in wood has been fundamental in establishing the main pathways of moisture ingress, its movement and the subsequent initiation of decay in any unprotected wood.

The internal rate of water movement in wood is influenced by the moisture diffusion coefficient whereas the surface emission coefficient governs the rate of transport between the surface of the wood and its surroundings,[1].

The movement of water through wood above the fibre

saturation point is complex because of the great variability in capillary flow through and particularly between wood cells. The source of moisture and its subsequent movement of flow above the fibre saturation point largely depends on the utilization of wood in service. A fence post or transmission pole will be wet where it is in ground contact and water is lost by evaporation above the groundline. Under these circumstances the wood was considered to be acting as a wick,[3] which subsequently creates conditions for decay, particularly at the groundline. The wick or mass flow phenomenon has been neatly demonstrated under laboratory conditions,[4], and the wick action is seen in Figure 1.

It has been shown that at least half the window frames examined in a regional survey in the United Kingdom in the late 1960's had at least one lower joint moist enough for fungal decay to occur,[5]. Specific points for water ingress were recognised as being through open joints and this is thought to be the principle cause for premature decay in window joinery.

Secondary Treatments with Solid Water Diffusible Preservatives

The understanding of the principles of water ingress in wood and the later realisation that water was also redistributed has been fundamentally important in the development of solid water diffusible chemicals as a secondary treatment. A basic theory and principle has now been established for this type of secondary treatment. If a solid water diffusible preservative is placed in the water flow or pathway it will be dissolved and distributed with the water. The water acting as an internal solvent is converted into a preservative solution which will reduce or eliminate the risk of fungal attack which would normally be created by the very presence of the water (see Figure 1). It is not always necessary to have mass flow, as under steady state conditions at appropriate moisture contents suitably formulated products will dissolve and be distributed by natural diffusion in the wood.

The attention focused on secondary treatments during the last ten years has probably paralleled the development and use of solid water diffusible preservatives such as fused rods consisting of inorganic borates.

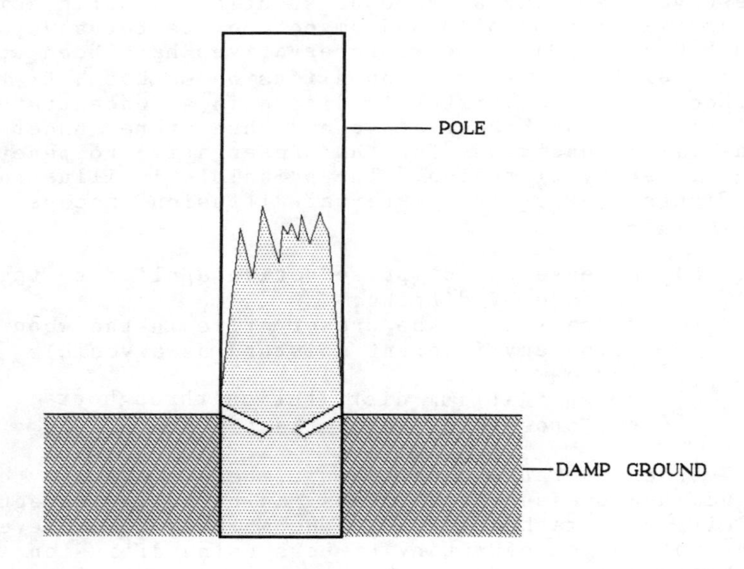

Figure 1 Mass flow of water or wick action in standing pole showing strategic placement of solid diffusible wood preservative as a secondary treatment.

2 SECONDARY TREATMENTS WITH BORATES

The reason for selecting the versatile borates as secondary treatment preservatives has already been discussed. Boron, as the active ingredient in the inorganic borates, has been used in all areas of wood preservation for many years,[6,7]. Several general reviews on their use have been published over the years,[8-10] with further more recent publications,[11,12]. The major use of boron has been as a pretreatment preservative using an aqueous solution of boric acid/ borax mixtures or of disodium octaborate tetrahydrate ($Na_2B_8O_{13}\cdot 4H_2O$). These preservatives have been applied almost exclusively by boron diffusion methods. By this method unseasoned timber is dipped in a concentrated solution of the preservative and then stored under non-drying conditions for the preservative to penetrate the timber by diffusion. The principle is illustrated in Figure 2, where the external diffusion process achieves:-

a) An envelope of preservative applied to the surface by dipping.
b) Migration of the preservative in the wood using any inherent moisture as a vehicle for osmosis.
c) An equilibrium distribution throughout the cross-section of the timber.

Under suitable conditions timbers resistant to vacuum and pressure treatments can be fully impregnated by diffusion methods,[13-19]. The ability to achieve good penetration in refractory timbers using diffusion technology is, perhaps, the greatest potential asset of this system.

This asset was also a major factor in selecting borates for use in secondary treatments. A similar principle is achieved "in reverse" by the fused borate rods using an internal diffusion process as seen in Figure 2:-

a) Rods are located within the wood in areas of high risk where there is moisture.
b) Migration of the preservative in the wood using the water in the high risk area.
c) An equilibrium distribution is achieved which is governed by the moisture regime in the wood.

Both external and internal diffusion processes are

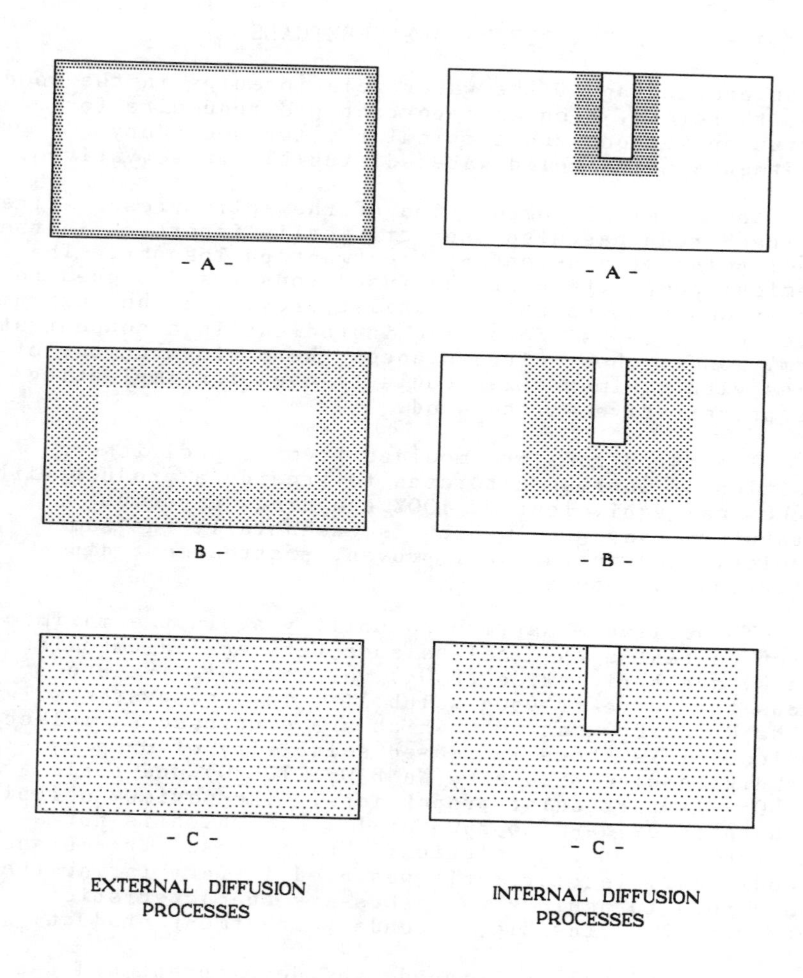

Figure 2 External and internal diffusion processes showing how equilibrium distribution of chemical is achieved with time.

dependent on time and moisture content in the wood,[15,20].

3 SOLID WATER DIFFUSIBLE CHEMICALS

An understanding of the water relationships in the wood has, therefore, been an important pre-requisite for diffusion methods and particularly for secondary treatments using solid water diffusible preservatives.

The chemical composition of the solid preservative or fused rods has also been critically important in the development of a secondary treatment philosophy. The chemical composition of the fused rods was designed to achieve maximum solubility whilst providing the maximum level of boron as the active ingredient in a concentrated form. Under these circumstances the minimum number of holes with minimum sizes could be used for implanting the preservative in the wood.

The fused rods are moulded from a specific composition of inorganic borates to provide a product with a chemical equivalent of 100% anhydrous disodium octaborate ($Na_2B_8O_{13}$) which is essentially the same chemical used in the well proven, pretreatment dip diffusion processes.

To achieve a maximum solubility against a maximum Boron content a chemical balance between sodium oxide (Na_2O) and boric oxide (B_2O_3) was used,[21]. Figure 3 shows the curve of B_2O_3 solubility in a saturated solution of Na_2O and B_2O_3 at 30°C which are expressed as molecular ratios. This shows increased solubility of B_2O_3 by blending sodium biborate ($Na_2B_4O_7 \cdot 10H_2O$) and boric acid (H_3BO_3) in particular proportions. The maximum solubility is a ratio of Na_2O to B_2O_3 at 1 : 3.8 which is not a particularly stable solution. Under these circumstances a modified molecular ratio was used in formulating the fused rods to achieve the necessary characteristics for a water diffusible, secondary treatment product.

As previously mentioned, the development and use of the fused rods has mirrored the greater awareness for the need for secondary treatments. This has certainly been the situation within the railway and transmission industries.

Secondary Treatments in Wooden Transmission Poles

The use of fused rods has been assessed as a

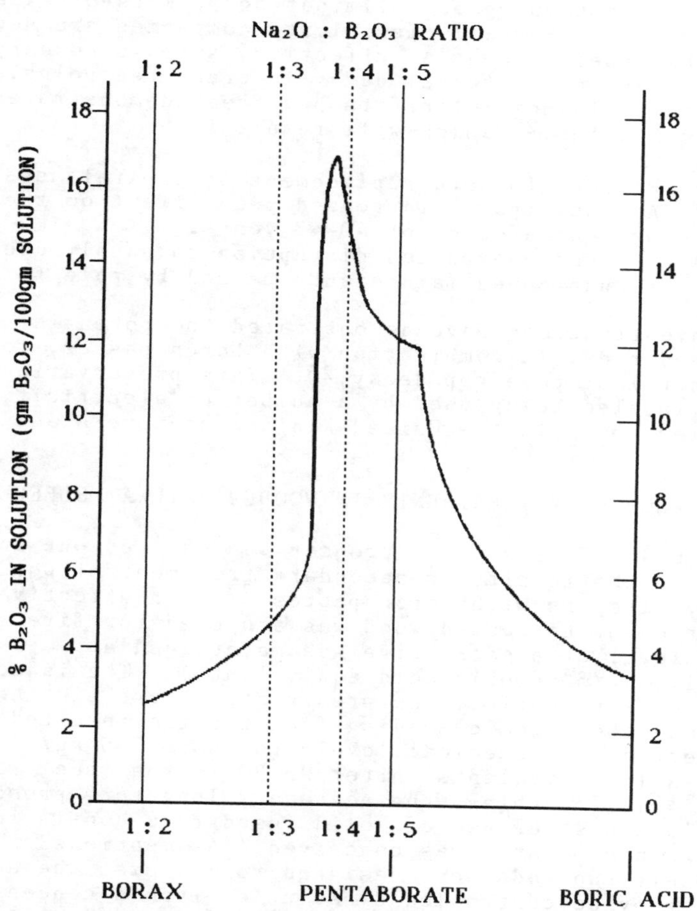

Figure 3 The solubility of boric oxide in water at 30°C when different molecular ratios of sodium oxide to boric oxide are present in the solution (From data supplied by Dominion Laboratory, D.S.I.R., Wellington)

secondary treatment for wooden transmission poles in the United Kingdom and Sweden,[22-24]. In the United Kingdom the results indicated that not only did the fused rods prevent colonisation by decay fungi but they were also capable of containing and eliminating pockets of existing infection. Several electricity companies are now using the fused rods as an effective, safe, secondary treatment against internal decay in creosoted poles. Significant financial benefits have been demonstrated to the companies,[25] which will result in:-

1) Reduced forward replacement of capital assets.
2) An anticipated increased pole life from the present average of 40-45 years.
3) Reduced unexpected disruption to supply due to unplanned maintenance caused by pole failure.

Investigations have demonstrated the potential of the fused rods, in combination with boron pastes, for the control of pole top decay,[25]. This preservation system is also being used by a number of electricity companies, both in the United Kingdom and overseas.

4 SECONDARY TREATMENTS IN WOODEN RAILWAY SLEEPERS

In the mid 1970's a major project was carried out in Sweden on the in-situ or secondary treatment of wooden railway sleepers with boron pastes,[27]. An effective protection of untreated wood was achieved for five years if the level of preservative was maintained at 1.2 - 1.5 kg/m³ (boric acid equivalents). The injection method of application, the preservative level in the paste and the dynamic loading from traffic resulted in a decrease of chemical levels to about 0.5 kg/m³ (boric acid equivalents) after 25-30 months (see Figure 4). To achieve the necessary long term preservative loadings the use of solid fused rods consisting of inorganic borates was conceived. As mentioned previously the rods were designed to maximise the active boron, provide controlled dissolution from a concentrated reserve and be easily inserted into the sleepers.

In the mid 1980's a further major investigation was carried out in the United Kingdom to develop a new maintenance system for wooden sleepers based on the use of the fused rods,[28,29]. Unlike the Swedish investigation the main water ingress and subsequent decay activity was found to be along the central longitudinal axis of the sleepers. Rods were placed along the centre of the

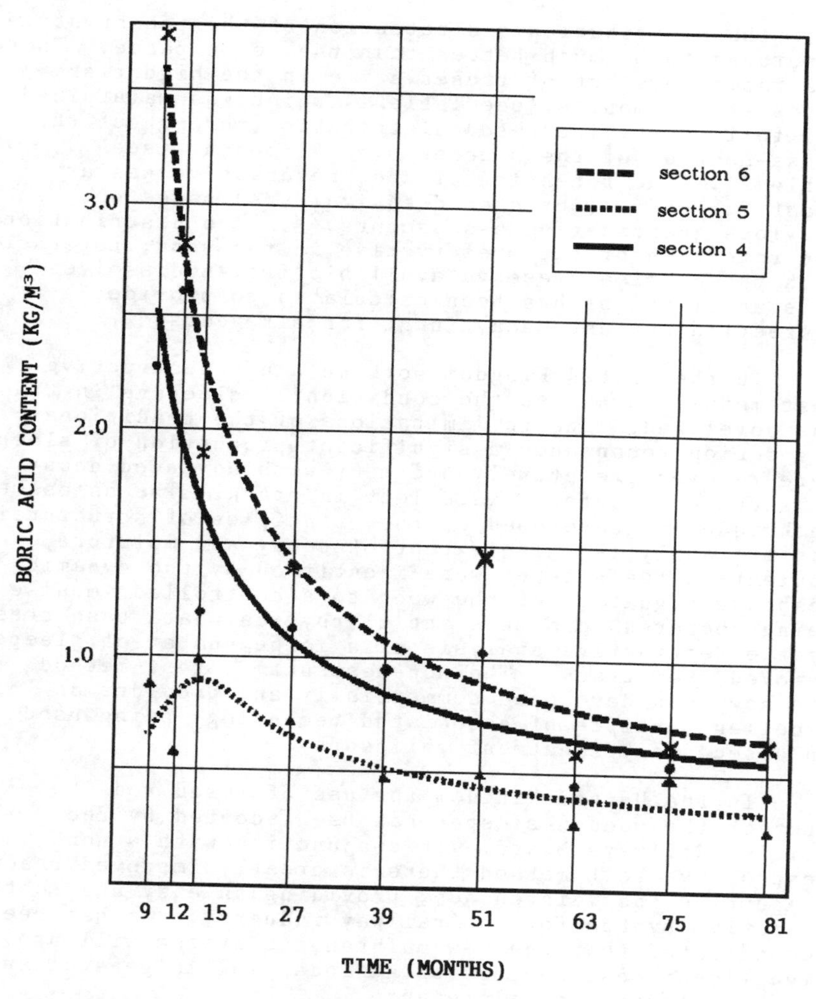

Figure 4 Graphs showing horizontal diffusion of boric acid in Swedish tests. Preservative profile mainly due to boron paste being lost from injection holes. (from Bechgaard et al, 1979)

sleeper and the distribution of the preservative was assessed over a number of years.

The distribution and retention of the preservative was found to be much better than the boron paste. There was rapid movement of preservative in the high risk areas at six months (see Table 1) which was maintained and followed by increased distribution throughout the cross-section of the sleeper over 51 months (see Table 2). The retention of the preservative was at least 2.5 kg/m^3 (boric acid equivalents) based on previous quantitative assessments,[29]. The distribution, and retention of the preservative in the wood, together with the chemical reserve available from undissolved rods at 51 months has been calculated to provide protection against decay fungi for six years.

In the United Kingdom work on a non-destructive test method to assess the condition of sleepers in track was developed. Due to limitations of the traditional inspection techniques a significant proportion of sleepers were removed prematurely and many with advanced decay or structural defects were left in track. The assessment technique was developed on the principles of structural dynamic analysis. The extent of decay and moisture content in the sleeper were identified by the dynamic response signature of the wood to a controlled impulse being compared to a substantial physical data base created by the destructive analysis of a large number of sleepers removed from track. The non-destructive test method has now been developed commercially and provides an excellent example of associated technology being used in a secondary treatment philosophy.

In the United Kingdom the use of fused rods to protect the wooden sleeper has been adopted by the British Railways Board. In conjunction with a non-destructive test method there is greatly improved track inspection and maintenance, providing an easy and cost effective system for the railway industry. It has been demonstrated that the new maintenance system will provide significant cost savings of at least 15% in general and patch resleepering maintenance.

It was also realised that fused rods may have a potential in the in-situ treatment of decay in joinery,[30] and this led to further investigations in the use of the product,[31,32]. This has seen the development of successful remedial treatments in joinery,[33]. Fused rods have also been used in specialised remedial treatment

Table 1 Distribution of Preservative from Fused Rods in Treating Decay Susceptible Timber in Centre Zone of Sleeper

TREATMENT PERIOD (Months)	Sleeper Sections (% Susceptible timber treated*)			
	1-6	1+6	2+5	3+4
6	58.9	47.5	40.1	82.3
12	42.3	28.3	17.3	76.0
18	51.5	64.5	17.6	78.9
51	62.8	63.3	44.9	80.0

(mean of 2/4 replicas)

*100 x No.Reagent 2/3, Moisture 1/2, Creosote 0/1 in blocks 21-35(inc)
____
Moisture 1/2, Creosote 0/1 in blocks 21-35(inc)

Each sleeper was sampled by removing six slices which represent the furthest points away from the rod positions. One slice was removed 200 mm from each end (sections 1 and 6); one under the centre of each base plate (sections 2 and 5) and one halfway between the first two rods inside the baseplates (sections 3 and 4).

Table 2 Distribution of Preservative from Fused Rods in Treating all the Timber in Sleeper Cross Section

TREATMENT PERIOD (Months)	Sleeper Sections (% all timber treated*)			
	1-6**	1+6	2+5	3+4
6	14.6	13.0	7.9	23.2
12	13.3	10.5	8.4	21.1
18	20.2	20.7	13.6	26.1
51	25.9	19.1	17.5	41.4

(mean of 2/4 replicas)

* $\frac{100 \times \text{No. Reagent } 2/3}{110}$

** $\frac{100 \times \text{Reagent } 2/3}{330}$

Each sleeper was sampled by removing six slices which represent the furthest points away from the rod positions. One slice was removed 200 mm from each end (sections 1 and 6); one under the centre of each base plate (sections 2 and 5) and one halfway between the first two rods inside the baseplates (sections 3 and 4).

operations as well as treatment of various timbers of architectural and conservation importance.

5 SUMMARY AND CONCLUSIONS

During the past decade solid water diffusible preservatives have been successfully applied as effective secondary treatments. They can be inserted into potential high risk areas where water ingress is possible or will occur and cause fungal decay problems. The use of fused rods enables an internal, moisture regulated preservation system to be strategically built into any wooden commodity where it is most required.

There is now no doubt that secondary treatments have an increasingly important part to play in the wood preservation industry. Their selected use as a supplementary, and indeed complementary, treatment will extend the service life of many wooden commodities and optimize the utilization of forest resources.

REFERENCES

1. C. Skaar, 'The Chemistry of Solid Wood', American Chemical Society, Washington, 1984, Advances in Chemistry Series, 207, Chapter 5, p. 45.
2. C. Skaar, 'Water in Wood', Syracuse Univ. Press, Syracuse, 1972.
3. J.F. Levy, Rec. Ann. Conv. Brit. Wood Preserv. Assoc., 1962, 3.
4. E.F. Baines and J.F. Levy, J. Inst. Wood Sci., 1979, 8 (3), 109.
5. C.H. Tack, Building, 1968, 214, 135.
6. E. Bateman and R.H. Baechler, Proc. Ann. Meeting AWPA, 1937, 33, 91.
7. J.E. Cummins, Australian Timber Journal, 1938, 4, 661.
8. D.R. Carr, Rec. Ann. Conv. Brit. Wood Preserv. Assoc., 1959, 1.
9. R. Bunn, N.Z. Forest Services, 1974, Tech. Paper No. 60, p. 112.
10. R. Cockcroft and J.F. Levy, J. Inst. Wood Sci., 1973, 6 (3), 28.
11. D.J. Dickinson and R.J. Murphy, Rec. Ann. Conv. Brit. Wood Preserv. Assoc., 1989, 35.
12. P. Vinden, First Int. Conf. on Wood Protection with Diffusible Preservative, Nashville. Proceedings 47355, Forest Product Research Society, 1990, 22.

13. R.H. Baechler and H.G. Roth, Forest Products Journal, 1964, 14 (4), 171.
14. J. Thornton, Borax Consolidated Limited, London, 1964, TP98.8/64, p. 19.
15. D.N. Smith and A.I. Williams, Timberlab Paper No. 5, Building Research Establishment, Princes Risborough Laboratory, 1969, p 11.
16. D.C. Markstrom, L.A. Mueller and L.R. Gjovik, Forest Products Journal, 1970, 20 (12), 17.
17. C.R. Levy, S.J. Colwell and K.A. Garbutt, Inter. Res. Group Wood Pres., 1972, Document No. IRG/WP/310, p. 67.
18. P. Vinden, Inter Res. Group Wood Pres., 1984, Document No. IRG/WP/3291, p. 18.
19. R.J. Murphy and D.J. Dickinson, Rec. Ann. Conv. Brit. Wood Preserv. Assoc., 1986, 46.
20. J.J. Morrell, C.M. Sexton and A.F. Preston, Forest Products Journal, 1990, 40 (4), 37.
21. K.M. Harrow, N.Z. Timber Journal and Forestry Review, 1955, 2 (3 and 4), 20.
22. D.J. Dickinson, P.I. Morris and B. Calver, Inter. Res. Group Wood Pres., 1988, Document No. IRG/WP/3518, p. 3.
23. B. Henningsson, H. Friis-Hansen, A. Kaarik and E. Edlund, Inter. Res. Group Wood Pres., 1986, Document No. IRG/WP/3388, p. 21.
24. D.J. Dickinson, P.I. Morris and B. Calver, Distribution Developments, 1989, 1, 9.
25. Preservation of Distribution Poles Using the Timbor Rod, Bio-Kil Chemicals Limited, 1991.
26. D. Dirol and J.D. Guder, Inter. Res. Group Wood Pres., 1989, Document No. IRG/WP/3518, p. 5.
27. C. Bechgaard, L. Borup, B. Henningsson and J. Jermer, Swedish Wood Preservation Institute, 1979, Report No. 135E, p. 62.
28. W. Beauford and P.I. Morris, Inter.Res. Group Wood Pres., 1986, Document No. IRG/WP/3392, p.13.
29. W. Beauford, P.I. Morris, A.M. Brown and D.J. Dickinson, Inter. Res. Group Wood Pres., 1988, Document No. IRG/WP/3492, p. 13.
30. D.J. Dickinson, Inter. Res. Group Wood Pres., 1980, Document No. IRG/WP/3159, p. 3.
31. J.K. Carey and A.F. Bravery, Inter. Res. Group Wood Pres., 1987, Document No. IRG/WP/2291, p. 12.
32. M.G. Dietz and E.L. Schmidt, Forest Products Journal, 1988, 38 (5), 9.
33. P.E. Dicker, D.J. Dickinson, M.L. Edlund and B. Henningsson, Rec. Ann. Conv. Brit. Wood Preserv. Assoc., 1983, 73.

Organic Solvent Preservatives: Application and Composition

E.A. Hilditch

CUPRINOL LTD., FROME, SOMERSET BA11 1NL, UK

1. DEFINITION

An organic solvent (O/S) wood preservative is one in which wood preserving chemicals are dissolved in a nonpolar organic solvent. The solvent may be volatile or nonvolatile.

Products using a volatile solvent are the most common and where used without qualification the term usually refers to this sort, they may be known as light organic solvent preservers (LOSP) or paintable preservatives. Products with a nonvolatile solvent are referred to as heavy oil or nonpaintable.

LOSP are far the most widely used, except where stated, this paper relates to them.

2. COMPOSITION

In the simplest form the only ingredients are the active wood preserving component and solvent, often however a combination of active ingredients is used, such mixtures may have a better profile of activity or some other advantage over single ingredients.

For some active ingredient and solvent combinations other components must be included to, for example, increase solubility or prevent formation of crystals on the wood surface

In addition to the necessary ingredients of a basic preservative other materials may be added to improve or extend the performance or usefulness of the product. The two most common are colour and water repellency.

Active Ingredients

Wood decay is due to attack by fungi or insects, the active ingredient must be able to prevent this attack continuously from the time of treatment to the end of the life of the treated component, commonly it is failure of the preservative that precedes failure of the component. Beyond this the active ingredient must be suitable for formulation in this type of preservative with other appropriate properties. (See below)

Taking the many requirements together only a few chemicals have ever achieved wide use in O/S wood preservatives, those in current or recent use are :-

	Typical concentration[1]
Fungicides	
Pentachlorophenol (PCP)	5 %
Tributyltin oxide (TBTO)	1 %
Copper Naphthenate	2.75 % Cu.
Zinc Naphthenate	2-3 % Zn.
* Acypetacs-copper	17 % (2.75 % Cu)
* Acypetacs-zinc	14.5 % (3 % Zn)
* Trihexylene glycol biborate	1 % (boric acid)
* (Azaconazol)	
Insecticides	
Lindane	0.5-1 %
Dieldrin !	0.5-1 %
*Permethrin	0.1-0.2 %
*Cypermethrin	0.05-0.1 %
*(Deltamethrin)	

* New coming into use in the last ten years.
() not yet significant in the U.K.
! Now prohibited in the U.K.

Solvents

Most organic solvent preservatives, especially in the U.K. and Europe, use a volatile solvent which evaporates after application to leave a dry, usually paintable surface. (LOSP)

Most usually the solvent is an aliphatic petroleum solvent of the kerosine type, with boiling range 140-270 °C.[2,3] For a quicker drying preservative white spirit (boiling range 140-200 °C) may be used.

For some uses, mostly where there is direct competition with creosote, a heavy, nonvolatile solvent is preferred, the boiling range of the solvent is typically 180-360 °C.[4,5]

Cosolvents and Antiblooming Agents

Some active ingredients, notably PCP are not sufficiently soluble in cheaper solvents such as kerosine and white spirit. Formulation must therefore contain an auxiliary material to increase the solvency of the system, typical of those used are trixylenyl phosphate, dioctylphthalate and diacetone alcohol.

Crystalline active ingredients such as PCP and lindane may, on drying migrate to and crystalise on the surface. Some cosolvents such as trixylenyl phosphate also prevent blooming (typically 5% PCP in kerosine will need 5-7% trixylenyl phosphate to act both as cosolvent and antibloom) but with volatile cosolvents a separate antiblooming agent must be added, commonly a resin, wood, petroleum and coumarone-indene resins are all used.

Use of a combination of active ingredients, can avoid the cost of a separate cosolvent or antiblooming agent, for example with zinc naphthenate plus PCP, the zinc naphthenate acts as both cosolvent and antibloom while in a mixture of PCP and TBTO the lower concentration of PCP needed (1.75 %) is within it's solubility in kerosine.

Water Repellents and Colorants.

The commonest water repellent is paraffin wax at between 0.5 and 2.0 %. For products to be over painted a resin is included to reduce the effect of the wax on drying, gloss and adhesion. Petroleum or coumarone resins are commonest although many types may be used.[6]

Wax water repellents have been used for around 50 years, there have been no great changes in technology despite shortcomings. They are initially very effective but if exposed to the weather unpainted are durable for less than 1 year, they are not ideal for painting, with oil paints there is some retardation of dry and loss of gloss if preservative absorption is high. Application and adhesion problems, varying in degree with the paint and application level but often severe, are found with water based paints.

Silicones and other materials that have replaced wax in many other fields are too expensive for wood preservation. Many also have major deficiencies, silicones for example being even more difficult than wax to over paint.

The simplest colours to formulate, with least effect on the other properties, are soluble dyes and bitumen. Solvent soluble dyes all have poor weather fastness, consequently bitumens dominate, the major shortcoming is the limited colour range.

A wide colour range, fast to weather can be obtained by use of fast pigments. Small amounts of very finely dispersed pigments are used in some organic solvent preservatives but for better decorative effect larger amounts are needed, these then require binders, suspending agents and other ingredients. Such ingredients alter the nature of the product, it becomes a wood stain, these are not considered in this paper.

3. PROPERTIES AND PERFORMANCE

The properties associated with organic solvent preservatives, as a class derive mainly from the solvent while the performance as a preservative depends mainly on the active ingredients.

Properties

The key features of organic solvent wood preservers are ready penetration and the absence of dimensional change in wood.

These differentiate O/S from water-borne. Ease of penetration enables treatments such as brushing or spraying to give sufficient preservation for many purposes, especially for remedial treatment in buildings and for DIY use. Freedom from dimensional effect leads to use for pretreatment of dimensioned timbers especially windows and other joinery.

Other important properties are, reseasoning not required, absence of offensive residual odour, rapid drying, clean and colourless* and the ability to overpaint*, varnish*, stain* or polish*. (*Unless negated by active ingredient.)

Cleanliness, drying, lack of permanent odour and overpaintability allow use on and in dwellings and the like, where these factors make creosote unacceptable.

As with all preservatives the components used in organic solvent wood preservatives are such that they do not have any deleterious effect on wood, they are non-deliquescent and odourless, do not reduce strength nor, when dry do they increase combustibility or give off noxious fumes in a fire.

Most LOSP do not affect working or surface feel nor do they have any other adverse effect on use, in particular treated timber, when dry can be painted, polished, stained, varnished, glued etc. There is no increase in corrosion of metal fittings. When dry they do not stain or otherwise harm plastic fittings, bleed into plaster, or anything in contact with the treated wood. However inclusion of water repellents, solvents of low volatility or solvent soluble colouring materials may necessitate special treatment where these are considerations.

O/S preservatives are the most expensive of the wood preservatives (per litre) but a simple comparison is misleading in relation to cost of treated wood. Rates of use of different preservatives for different purposes differ as do capital cost of plant and it's utilisation. Generally processing treatment costs are highest with water-borne especially if it must be subject to accelerated fixation or reseasoning. None the less total costs are generally highest with organic solvent preservatives, consequently their main use is where their advantages, particularly freedom from any distorting effects, or the ability to use simple methods of application are a major consideration and where adequate preservation results from limited fluid use.

Performance

The ability of any preservative to prevent decay or insect attack depends primarily on the chemical or combination of chemicals making up the active ingredient. The solvent or other ingredients might influence efficacy but only to an extent.

Wood is attacked by several hundred, perhaps thousands of fungal species from most divisions and classes. Fungicidal effectiveness must therefore be broad and must include all fungi that decompose wood in any "economic" situation, wet and dry rot in buildings, fungi that are active in the soil and that decay posts, poles, etc. and for universal use aquatic fungi (marine and fresh).

In temperate climates only a few species of beetles and weevils attack timber in use. In the tropics the range of beetles is greater while termites are the major destroyers of wood. None the less insecticides used in O/S wood preservatives have a broad spectrum of activity. (All were initially introduced for other purposes.)

A wood preservative is expected to retain it's activity for many years, perhaps 50 to 100. The active ingredients are chemically stable, volatility and water solubility are extremely low. There are some differences between different chemicals, these tend to be reflected in chosen use.

The efficacy of a preservative treatment depends as much on the amount in the wood and on it's distribution in the wood, variation in these result from choice of treatment method. Pressure treatment or long soak in for example copper naphthenate penetrates deeply and will protect poles in the ground for 20-30 years in either volatile or nonvolatile solvent, PCP gives similar service in heavy oil but is less effective in a volatile solvent. Shorter immersion or double vacuum treatment, while penetrating less is still enough to give long service with any of the main preservatives to building components such as window frames, roof trusses etc. where the decay hazard, while sufficient to warrant treatment, is less than in the soil.

Inclusion of water repellents in a wood preservative reduces the rate at which liquid water is absorbed. In situations with alternate wetting by rain and drying this increases the time to get wet so that often drying starts before the wood has reached peak moisture content. The time for which the wood is wet enough to decay is then reduced, as is the rate and range of movement in the timber. The rate of loss of preservative due to leaching is also reduced.

4. USES

Organic solvent preservatives are used for the industrial treatment of timber prior to installation, for the remedial treatment of buildings and by the general public for both pretreatment and as a remedy.

Market (million litres per year)

	LOSP		Heavy oil
	U.K.	Europe	U.S.A.
Industrial	22	14	68
Remedial	4	?	?
DIY	10	30?	?
TOTAL	38		

Industrially in the U.K. about 45% of all treated timber is treated with an organic solvent preserver, usage is primarily for the pretreatment of building components, especially windows, doors, their frames and similar joinery. Application is by double vacuum for permeable softwoods and by double vacuum pressure for softwoods more resistant to treatment and for hardwoods or by immersion. Preferred products are those that have least effect on application or durability of paint, up to the present TBTO based formulations are most used for this reason, a change may be imminent caused by health and environments factors relating to organotin compounds.

Organic solvent preservatives are the main and most effective type used for in-situ remedial work in buildings. Application is mainly by spray with some injection. For many years this market was dominated by dual purpose products active against both insects and fungi and based on either PCP or TBTO with lindane. TBTO is now prohibited for this use in the U.K., lindane must not be used where bats roost and is now little used, use of PCP is minimal because of public opinion and proposed regulations. Increasingly solvent based products use trihexylene glycol borate, acypetacs zinc or other zinc soaps with permethrin or cypermethrin. For insect eradication treatment is moving to products that are specific for this purpose, because of solvent concerns, increasingly with aqueous emulsions.

LOSP are the only wood preservative type used domestically (DIY) for treatment of timbers in or for use in building. Much is also used on fences and other outdoor timbers in competition with creosote, today there is also competition for fence treatment with a range of dilute polymer emulsion colorants, these products do not offer any preservation against decay as do organic solvent preservatives and creosote.

In the U.S.A. transmission poles[7] may be treated with an organic solvent preservative in a heavy, nonvolatile solvent. Until recently the active ingredient was almost entirely PCP but now it is being replaced by copper naphthenate. Usage in the USA is around 68 million litres a year. Use in the U.K. is only on fence posts and is trivial.

For many applications sheds, fences, floor boards, feature beams etc. the user wants both to preserve and colour the wood. Products including colour are used for these purposes, both industrially and by amateurs.

5. APPLICATION

The nature of timber is such that liquids only penetrate slowly, with the simplest of all treatment methods, total immersion, penetration depends on the nature of the liquid and on immersion time, with an organic solvent preservative immersion for 3 minutes results in a penetration of around 6 mm in permeable timbers. Such penetration, with suitable active ingredients, gives sufficient protection for many purposes including some building components,

Deeper penetration can be achieved by extending the immersion time, but to double penetration (12 mm) the time must be increased to over 1 hour with 30 mm only being reached after about 1 day. None the less it is the relatively ready penetration that is a key feature of LOSP. Typically creosote penetrates 20 mm in 1 day but water only 5 mm.[8]

Industrially an extended treatment time is a major cost and inconvenience and where better protection is required different methods are preferred. The rate of penetration can be increased either by applying pressure to the preservative fluid or by applying a vacuum to the timber, so creating an internal vacuum to draw in preservative. Industrial practice is to use double vacuum or short immersion (3-10 minutes) for the pretreatment of joinery and building components.

Remedial treatments within buildings are restricted by what is practical, use is therefore mainly spray with injection for heavy infestation or very large timbers.

The other main user group for organic solvent preservatives is the householder, again, restricted to what is practical and what is economic for occasional, small users. Most application is by brush with some spray and with immersion (Up to 24 hours) for severe situations.

Double Vacuum and Pressure Processes

Within the U.K. treatments using organic solvent preservatives are standardised into double vacuum for European Redwood (*Pinus sylvestris*) and similar permeable woods and double vacuum pressure for less permeable timbers.

In the double vacuum process timber is charged into a vessel which is then closed and an initial vacuum of -0.33 bar drawn and held for 3 minutes. For double vacuum pressure process the vacuum is greater and the period longer. This draws air out of the wood so creating an internal vacuum.

The vessel is filled with preservative, then the vacuum is released. The resulting pressure gradient increases the rate of penetration. Atmospheric pressure held for 3 minutes.

With the double vacuum/pressure process a hydraulic pressure is applied to further increase the rate of penetration. For different treatments this ranges from 5 minutes to 1 hour at 1 bar or 15 minutes at 2 bar.

Preservative is then pumped out of the vessel and a second vacuum of -0.67 bar drawn and held for 20 minutes. This second vacuum draws out some of the fluid, perhaps 40%. Finally vacuum is released and timber discharged.

The combined effect is deep penetration with lower preservative usage than for similar penetration by other means. Draining and drying are faster. Economy is better. In prepared European redwood sapwood penetration is around 15 mm with final preservative retention 0.5-0.6 litres per square meter surface. This is about double that for 3 minutes immersion but around half the retention of an immersion treatment giving similar penetration.

Anomalously the double vacuum treatment is considered, to be the equivalent of 3 minute immersion for some timbers but the equivalent of 1 hour for others.[9]

Penetration and preservative absorption vary with changes in the three stages,[10] for specific components specifications lay down treatment schedules.[11,12]

Commonly treatment plant comprises a treatment vessel with rectangular section, 1 - 2 metres high and wide, 10 -20 metres long, a tank for working fluid , a storage tank, transfer pumps, vacuum and pressure pumps, valves and controls to enable the cycle to be gone through automatically, with bogies and tracks to load the timber. Mostly the timber is loaded and the door closed manually while the treatment cycle is automatic.

In the USA and to a minor extent elsewhere heavy oil solutions of pentachlorophenol or, increasingly copper naphthenate are applied to transmission and similar posts by full pressure processes, these are similar to the double vacuum process but using pressures up to 14 bars. There are consequently major differences in plant construction. Such processes achieve complete penetration of sapwood. Preservative absorption is up to 300 litres per cubic metre.

Immersion

Immersion requires little description, the timber is put in the preservative fluid and held submerged for the specified time. In industrial practice, equipment is designed for mechanical handling with automatic control. In the U.K. most treatments are for 3 minutes, if a more severe treatment is needed DV or DVP processes are used but longer times are common elsewhere and for farm, estate or household use.

Brush and Spray

For industrial use these are too expensive in labour and too subject to skimping. For professional remedial treatment, spraying and for amateur (DIY) use, brushing are the preferred methods, because of practicality. The brush or spray are simply means of transporting fluid onto the timber surface, preservative which then soaks in until it is all absorbed or runs off, repeated application will improve the effectiveness of the treatment. Application must be on all surfaces and while it should be fairly uniform the careful brushing of painting is unnecessary, simply put on until recommended loading is achieved or as much as is practical.

6. SAFETY AND ENVIRONMENTAL

Pesticides, including wood preservatives are at the forefront of public concern on the effects of chemicals on health and the environment.

From an environmental standpoint wood preservatives differ from other pesticides in that they are put into wood, the active ingredients are designed to persist in the wood for decades. They are not deliberately sprayed around or put onto food, environmental pollution or entry into the food chain only occurs as a result of accident or improper disposal of waste, not as part of their proper or normal use. These must be guarded against but they are common to all chemicals and many other things.

Even so health and environmental considerations are now paramount in relation to choice of active ingredients. Most countries have regulations regarding wood preservatives some, including the U.K.[13,14], require positive clearance before marketing. Many do not allow the use of Dieldrin while PCP, TBTO and lindane are partly restricted. Use of these ingredients remains substantial but is reducing as alternatives are developed.

To meet current requirements an active ingredient must have low acute and chronic mammalian toxicity, must not be mutagenic, teratogenic or fetotoxic. It must not accumulate in the body. It must not cause irritation nor be unpleasant in any way. Full toxicological test data must be produced in evidence.

No harm must result from contact during normal use and there must be a generous "margin of safety". The active ingredient must be fixed in the wood sufficiently not to expose users of the wood or of wooden buildings or articles to harmful amounts either through direct contact or via the air. Precautions to be followed in use are laid down by the authorities in giving consent to sell and use products. For industrial and remedial use supplementary guidance notes are issued.[15]

The manufacturing process must not result in toxic/polluting waste nor must the product contain dangerous impurities. Both manufacturing and treatment plants are expected to be more tightly controlled under environmental legislation.[16]

Accidental discharge is a major concern particularly with products including ingredients that are on the marine pollutants "Red list" (PCP, TBTO, Lindane).[17]

The solvent (in LOSP) evaporates to atmosphere. In the work place sufficient ventilation must be provided to keep below occupational exposure standards.[18] For white spirit this is 100 ppm or 575 mg m^{-3}. Factories producing organic solvent preservatives, with modern, properly positioned air extraction can generally stay around 1-2% of this figure only exceptionally exceeding 5%.

Discharge of solvent to the general atmosphere is now of concern and reduction is sought. Photodegradation of hydrocarbon solvents is rapid and while it is unlikely that they contribute significantly to the greenhouse effect, they contribute somewhat to the formation of ozone in the lower atmosphere and to photochemical smog . The significance of this varies from place to place, problems arise where there are temperature inversion effects as in Los Angeles but are uncommon in the U.K. None the less an overall reduction in solvent discharges is being sought, German environmental legislation imposes a discharge limit on paraffins of 150 mg m^{-3}.[19]

This is a key issue for the future of organic solvent preservatives. The total amount of solvent used in wood preservatives in the U.K. of around 30 thousand tonnes a year is only a small proportion of the total for all volatile organic compounds but the industry needs to convince authorities and the public that reductions in total discharge can be found elsewhere with less loss in benefits.

For industrial pretreatment it is technically possible to recover the solvent for recycling but it is expensive in both plant and running cost and is not currently practiced.

For remedial and DIY use there is no practical solution to this problem although some changes in practice may come in the future.[20]

7. ACTIVE INGREDIENT CHEMISTRY

Metal Carboxylates

Naphthenates. Copper naphthenate was identified as a potential wood preservative in the 1880's as part of early Russian work into the composition of petroleum and the use of it's various components. It was first used commercially in 1912 in Denmark under the trade name "Cuprinol". Copper naphthenate continues in use today so that, creosote apart, it has the longest history of any material now in use. Copper compounds are green, zinc naphthenate came into use soon after as a colourless treatment.

Naphthenic acids occur naturally in crude oil, the content varies considerably with source.

With older refining techniques these acids were by products that had to be extracted to produce good fuels. Prices were accordingly low. Currently these acids are destroyed by cracking so that there is less available.

Naphthenic acids are themselves moderately effective fungicides, work in the 1930's, partly on wood, partly on textiles, showed that carboxylates such as the stearate, oleate, linoleate, tallate, rosinate etc. were less effective.

Acypetacs. Branched chain acyclic saturated carboxylic acids produced by the OXO or Koch processes became available around 1960. Investigation, by Cuprinol, on acids (of both types) in the C_7–C_{10} range showed the copper and zinc salts to have similar fungicidal activity to the naphthenates but to be more readily leached out.

A few years later it was found that combinations of primary or secondary acids (OXO) with tertiary acids (Koch) retained activity better that either type alone and were closer to naphthenic in this respect.[21] Known as Acypetacs-zinc or copper these are now used for industrial remedial and amateur treatments widely in the U.K.

Composition and structure. Within this field are two matters that have only received limited attention and which should be of particular interest to the more academic chemist. These are the composition of naphthenic acids and the structure of the zinc salts.

Naphthenic acids, like many naturally occurring materials, contain many different specific chemicals, essentially they are mixed alkyl substituted cyclopentane and cyclohexane carboxylic acids with the acid group in the side chain. Work on the composition was mostly done in the first half of this century,[22] techniques were involved and time consuming, modern techniques can be expected to give information on minor components or on variations between acids from different fields or in different fractions.

Zinc forms both normal and basic salts with carboxylic acids, solutions of zinc with naphthenic acid can be prepared with molecular ratios from 1:2 (normal salt) to, perhaps over 1:1. Viscosity measurements, of solutions of equal zinc content, show a minima at 1.3:2. This corresponds to a compound $Zn_4(Acid)_6O$ and may indicate a structure as has been suggested for beryllium oxide acetate.[23] With larger zinc atom and acid molecule (say C_{20}, acid value 180) there may be steric problems, if there are what is the structure? If not the molecule is large with molecular weight over 2000. I am not aware of any specific work on the basic salts of zinc with higher molecular carboxylic acids.

Organo metallic

Tributyltin oxide (TBTO). Bis (n-tributyltin) oxide came into use as a wood preservative in the 1960's[24] and is widely used for the treatment of window joinery in the U.K. It s main advantage for this use is the absence of any adverse effect on painting. TBTO does not perform well in soil contact and while it s adequacy in other situations, including joinery has been questioned established instances of failure are few.

The health and environmental hazards associated with TBTO are significant, it is a "Red List" substance for marine pollution and it's use in antifouling paints is now largely prohibited. Because of its potential effects on users the U.K. Health and Safety Executive have (1990) withdrawn approval of use in remedial and amateur products. Its use for industrial pretreatment is still under review.

Organo Chlorine

Pentachlorophenol (PCP). First use of PCP in wood preservation was around 1930, it achieved the widest use world wide. Use probably exceeded that for all other active ingredients and included all fields of wood preservation, pressure application to poles and structural engineering timbers, double vacuum of immersion to building timbers with spray, brush or other for remedial and DIY. It was widely used in wood stains.

PCP is a broad spectrum fungicide with bactericidal properties,

As a result of it s wide use PCP has become widely distributed in the environment.[25,26] Coupling this with questions on its effect on health has lead to a UN recommendation to reduce use. Some countries have restricted use in wood preservatives, restrictions in Europe were proposed in 1988,[27] debate on the justification for restrictions continue and the proposal has not so far been agreed.

Pentachlorophenol is undoubtedly a toxic material but there is uncertainty as to how much is attributable to the pentachlorophenol and how much to impurities particularly dioxins.

<u>Lindane.</u> Lindane is the pure (99%) gamma isomer of hexachlorocyclohexane (HCH), [28,29] it is a powerful insecticide with good persistence and is extensively used in wood preservatives for prevention or eradication of insect attack.

As a result of agricultural use, lindane has become widely distributed in the environment There is controversy around it's use. The controversy extends to it s use as a wood preservative, a specific concern in this connection is it's effect on bats. It is fatal to bats and must not be used where they roost. Beyond this the main environmental concern resulting from it s use as a wood preservative is for pollution of water by accidental discharge and it s effect on marine life. It is on the "red list" of marine pollutants.

Various effects on human health have been claimed but there is limited validation to many and while some countries and some specifiers restrict it s use it is still one of the most used insecticides in industrial preservatives.

<u>Dieldrin.</u> Several chlorinated hydrocarbon insecticides including dieldrin, aldrin and chlordane have been used in wood preservatives. Their use is now minimal because of concern about their health and environmental effects. Some are prohibited in several countries.

<u>Synthetic pyrethroids.</u>

Permethrin, cypermethrin and deltamethrin are related to naturally occurring pyrethrum. They are broad spectrum insecticides with low mammalian toxicity and have become widely used in place of lindane in wood preservatives.[30]

<u>Other actives.</u>

Historically chlorinated naphthalenes, either as mixture of mono and di compounds or of the polychloronaphthalenes were used. Ortho phenylphenol was and still is used to a limited extend.

Organo boron esters (trihexylene glycol biborate), are used in the U.K. for remedial treatment for a few years. The activity approximates to the equivalent amount of boric acid.

Ingredients that are used (not all in the U.K.) on a limited basis, mostly in special products include dichlorfluanid (N-dimethyl N'-phenyl-N'-(fluorodichloromethylthio)-sulphamide)), particularly where blue stain resistance is required, carbendazim (methyl benzimidazol-2-ylcarbamate), furmecyclox (2,5-dimethyl 3-(N-cyclohexyl-N-methoxy)furane carboxylic acid amide) and oxine-copper (Copper 8-hydroxyquinolinate).

As objections arise to use of the well established ingredients a number of materials have been developed, these include IPBC (3-iodo-2-propynyl butyl carbamate) and TCMTB (2-(thiocyanomethylthio)benzothiazole) azaconazole (1-[(2-(2,4-dichlorophenyl) 1,3-dioxolan 2-yl))methyl]-1 H-1,2,4-triazole) and the related tebuconazole and propiconazole. Many of these have a limited spectrum of activity and are best used in combinations. They have yet to find their proper place.

8. CONCLUSION

Organic solvent preservatives have been used for about 80 years, industrial and remedial use has been extensive for over 30 years yet they still present both challenge and opportunity for chemists.

Potentially they are the best of all the wood preservatives, they penetrate readily without detrimental effect. Correctly formulated they are effective in any use situation, can be applied by both simple and sophisticated techniques and are clean.

The challenge is to provide both solvents and active ingredients that both do these things and have less "down side".

The solvents, in most cases simply evaporate, those used are not very bad from a general health or environmental stand point but - they are bad, not good.

Many of the active ingredients that have been the core of the industry are now being phased out on health, environmental or other grounds only one or two have general acceptance, perhaps none have universal acceptance on all counts.

These problems cannot be solved by simply using an other type of preservatives, these have equal or greater problems, creosote will never be much better than it is now, it has it s own health and environmental shortcomings. Water will always cause wood to swell and will only penetrate slowly, current active ingredients like copper chrome arsenate share in full measure the health and environmental problems.

Avoidance of preservatives is also not a solution, use of unpreserved softwoods will mean more frequent maintenance and replacement, expensive and needing more trees, use of durable hardwoods in the immediate future means tropical hardwoods. Using materials other than wood, all entail environmental considerations, many of the substitutes do not have the renewability of wood.

So the challenge is for improved solvents - perhaps partly achievable through simple changes to refining techniques, partly most difficult - and for active ingredients meeting tomorrows standards requirements for effective performance coupled with safe use and absence of environmental effect.

The immediate challenge is for the industry in accepting the need for a sufficient level of research to carry the industry forward and allow the public to continue to benefit from the preservation of wood.

9. REFERENCES

1 British Wood Preserving Association Manual 2.0 Specification for Biocides: 1986
2 British Standard BS 5707:Part 1:1979. Solutions of Wood Preservatives in organic solvent. Part 1 Specification for solutions for general purpose
3 American Wood-Preservers Association Standard P9-87. Standards for Solvents and Formulations for Organic Preservative Systems.
4 British Standard BS 5707: part 2:1979 Solutions for Wood Preservatives in Organic Solvent. Part 2. Specification for pentachlorophenol Wood Preservative
5 American Wood Preserver's Association. Standard P9-87 Standards for Solvents and Formulations for Organic Preservative Systems.
6 Hilditch,E.A. The relative effectiveness of different materials and methods of treatment. Symposium on Water Repellents and Wood. Forest Products Research Laboratory Princes Risborough 4 May 1965.
7 American Wood Preserver's Association. Standard C4-90. Poles - Preservative Treatment by Pressure Processes. (1990)

8 Non-pressure methods of applying wood preservatives Forest products Record No. 31 H.M.S.O. 1961
9 British Standard BS 5589:199 Code of practice for Preservation of Timber
10 Purslow,D.F. Methods of Applying Wood Preservatives. Building Research Establishment Report. H.M.S.O. 1974
11 British Standard BS 5589:1989 Code of practice for Preservation of Timber
12 British Standard BS 5268: Part 5:1989 Structural Use of Timber. Part 5 Code of Practice for the Treatment of Structural Timber.
13 Food and Environmental Protection Act 1985. HMSO.
14 The Control of Pesticides Regulations 1986. Statutory Instrument 1986 No. 1510. HMSO.
15 Guidance Note GS46: In-situ treatments using timber preservatives. Health and Safety Executive. 1989.
16 Environmental Protection Act 1990. HMSO.
17 The Common Inheritance. Britain's Environmental Strategy HMSO. 1990.
18 Occupational Exposure Limits 1990. Guidance Note EH40/90 Health and Safety Executive.
19 Bundesanzeiger "TA Luft" . February 1986.
20 Hilditch,E.A. Preservative Treatments- Where are we now and Future Trends. JOCCA. 1990(10) 405-407 & 420.
21 Hilditch.E.A., Sparks.C.R., Worringham.J.H.M. Further Developments in Metallic Soap based Wood Preservatives. Annual Convention, British Wood Preserving Association 1983.
22 Lochte.H.L. Littmann.E.R. The Petroleum Acids and Bases. Constable. London. 1955.
23 Mehrotra.R.C. Bohra.R. Metal Carboxylates. Academic Press. London 1983.24 Richardson,B.A. Organotin Wood Preservatives. Activity and Safety in relation to Structure. Proceedings American Wood Preserver's Association 1988, 56-69
25 Environmental Health Criteria 71. Pentachlorophenol. Geneva: World Health Organisation. 1987
26 Rango Rao,K. (Ed.) Pentachlorophenol. Chemistry, Pharmacology and Environmental Toxicology. New York: Plenum Press. 1977
27 EEC 88/C/117/11. Draft amendment to Directive 67/548/EEC Official Journal C117. 4 May 1988
28 Specification WHO/SIT/3.R3 Lindane. World Health Authority 1965
29 Biegel,W. (Ed.) Lindane: Answers to Important Questions Brussels: Centre International d'Etudes du Lindane. 1988
30 Carter,S.W. The Use of Synthetic Pyrethroids as Wood Preservatives. Annual Convention, British Wood Preserving Association, 1984. 32-41

Wood–Chemical Interactions and Their Effect on Preservative Performance

Alan F. Preston and Lehong Jin

LAPORTE TIMBER DIVISION, RESEARCH AND DEVELOPMENT, ONE WOODLAWN GREEN, CHARLOTTE, NORTH CAROLINA 28217, USA

ABSTRACT

The role of fixation mechanisms of wood preservatives with various wood components in relation to biodeterioration processes is reviewed. This includes studies on the effect of fixation mechanisms of borates, quaternary ammonium compounds, chromated copper arsenate and other copper-based preservatives on distribution patterns and performance. The influence of these factors on the development of alternative preservatives is also addressed in the light of test methodologies and approval processes.

1 INTRODUCTION

The increasing pressures on the wood preservation industry from an environmental viewpoint has led to recent moves towards accelerated fixation procedures for waterborne preservatives,[1,2] concern over the ultimate disposal of treated wood, and improved treatment plant practices. Furthermore, moves towards the use of the most environmentally acceptable treatment for a particular application have given rise to new wood preservative standards based on the Hazard Class or Use Category systems. In response to the environmental concerns, research is now on-going in the development of new preservative formulations, the use of water repellents to enhance both product performance and environmental aspects[3,4] and into increased study of wood modification and biotechnology as alternatives to the application of biocides for wood protection.

In order to achieve many of these changes taking place in the wood preservation industry worldwide it is of value to understand the role of fixation mechanisms of wood preservatives with various wood components in relation to biodeterioration processes. Furthermore the influence of fixation mechanisms on distribution patterns and performance should be understood in order to improve the performance of current products and to develop new systems. This paper addresses these aspects as well as their influence on the development of alternative preservatives.

2 CHEMICAL NATURE OF WOOD

Wood is comprised primarily of cellulose (40-50%) and hemicellulose (20-35%) (known collectively as holocellulose), and lignin (15-35%). There is a minor amount of extraneous materials (2-10%) in wood, mostly in the form of organic extractives such as tannins, lignan, flavonoids, stilbenes, terpenoid, starch, lipids, pectins, alkaloids, proteins, fat and waxes and as well as trace amount of inorganic minerals (0.1-1.0%).

Cellulose is the most important component in wood and it consists of 1,4-ß-linked glucopyranose sugar units having both intermolecular and intramolecular hydrogen bonding. The average cellulose chain length (or degree of polymerization) is in the 7,000 to 10,000 glucose units range. Cellulose consists of a crystalline area where cellulose chains are arranged in an orderly three dimensional crystal lattice as well as amorphous regions where the cellulose chains show much less orientation with respect to each other.

Hemicelluloses are complex mixtures of short chain polysaccharides which are either water or alkali soluble. The chemistry of hemicelluloses has been widely studied and has been found to be much more complex than that of cellulose. In softwoods the galactoglucomannans and arabinoglucuronxylans predominate, while in hardwoods the glucuronoxylans and glucomannans are the most frequently occurring structures.

The third major wood component, lignin, is a random three dimensional polymer of three basic phenylpropane monomers. The phenylpropane units are linked by biphenyl, aryl-alkyl or ether linkages, and form relatively stable and inactive polymers that are resistant to hydrolysis. Lignin can be divided into several classes according to their structural elements.

Guaiacyl lignin, which occurs in almost all softwoods, is largely a polymerization product of coniferyl alcohol. The guaiacyl-syringyl lignin, typical of hardwoods, is a copolymer of coniferyl and sinapyl alcohols, the ratio varying from 4:1 to 1:2 for the two monomeric units. The concentration of lignin is higher in the middle lamella than in the secondary wall. However, because of the thickness of the secondary wall, at least 70% of the lignin in softwoods is located in this region.

Lignin has been proven to be a major fixation site for certain wood preservative components,[5-8] such as copper. The level, type and location of lignin within the wood structure have a significant impact on the fixation of wood preservatives and thus the microbial susceptibility of the wood.

3 WOOD BIODETERIOGEN INTERACTIONS WITH WOOD COMPONENTS

The organisms primarily responsible for the biodeterioration of wood are the members of the Basidiomycetes group, and these are usually classified as either brown rot or white rot fungi. A third group of fungi known to cause significant breakdown of wood components are members of the Fungi Imperfecti known as soft rot fungi. Other organisms such as staining fungi, molds, bacteria and Actinomycetes also effect various wood properties but do not usually cause failure of the wood structure.

The brown rot fungi primarily degrade cellulose leaving the wood a brown color attributable to the predominance of the residual lignin. Both enzymatic and chemical mechanisms of action have been shown to be operative with brown rot fungi. White rot fungi attack both lignin and cellulose and can cause catastrophic failure similar to that seen with brown rot organisms. Soft rot decay occurs on both the cellulose and lignin components of wood but at a much slower rate than is usually the case with white rot decay.

4 CCA AND ITS INTERACTIONS WITH WOOD

Since its development in India in 1933, CCA has grown to become the dominant wood preservative worldwide in terms of volume of wood treated. While initially this growth was slow, in the past thirty years the use of CCA has grown rapidly. In the U.S. this growth was particularly marked during the last decade. CCA has become widely accepted because it provides a treated product which is odorless and clean to handle, while retaining the necessary attributes of longevity in the protection of wood from biodeterioration. CCA was the first chemically fixed preservative to find widespread usage where predecessor products such as creosote and pentachlorophenol had relied on relative insolubility to allow them to reside in wood for prolonged periods. The fixed nature of CCA also opened the way for the use of wood preservatives in domestic construction applications and the consequent growth of the do-it-yourself market for treated lumber.

The fixation of CCA in wood has been studied by a number of research groups[9-17] and while there is general agreement on the key features of the fixation process of CCA in softwood, the mechanisms involved are complicated and not fully elucidated even now, nearly sixty years after the development of prototype CCA systems.

The initial reactions of CCA with wood involve complexation of Cr(VI) with lignin followed by reduction of the chromium to Cr(III). Copper also reacts directly with wood, by ion exchange, and then a series of condensation reactions occur to give chromium arsenate, copper arsenate, copper chromium, and perhaps copper chromium arsenate complexes. While some of these may be associated with wood components through coordination or covalent complexes, others are fixed by insolubilization within the wood structure. The formation of complexes between arsenate and wood components is possible but undocumented.

Experiences from field tests and practice have shown that the retention of CCA necessary to control decay, particularly soft rot attack, is much higher in many hardwood species than that with softwood species.[18-19] Uneven microdistribution of CCA in cell wall[20-21] which largely relates to lignin level, type and location[23-24] has been suggested as one of the causes for the poor performance against soft rot in hardwoods. As mentioned earlier, the type and amount of lignin present also influence the CCA performance in hardwoods. The need for higher preservative retentions in hardwoods than in softwoods has also been observed with other preservatives systems[25].

5 NEW PRESERVATIVES - PRINCIPLES FOR DEVELOPMENT

The current major wood preservatives used worldwide,

namely creosote, pentachlorophenol, chromated copper arsenate and ammoniacal copper arsenate (now ammoniacal copper zinc arsenate) were all introduced over fifty years ago and in the case of creosote, over 150 years ago. In the face of the environmental pressures dating back to the 1970's, there has been increasing interest in the development of wood preservatives with lower perceived environmental drawbacks than exists with the current formulations.

In order for wood preservatives to perform the functions of long term protection of wood from a multiplicity of biodeteriogens, they must be broad spectrum, stable in wood for a long service period, and soluble in a carrier to allow initial treatment of the wood. Unfortunately, these requirements run counter to biocide developments in the area of agrochemicals over the last thirty years, where the criteria for development has been towards chemicals which are organism specific, readily biodegradable in the environment, and insoluble to allow attachment to the target substrate by surface deposition. These conflicting needs, coupled with the increasing disparity between the chemical cost of the existing wood preservatives and modern organic agrochemicals, has meant that the more recently developed agrochemicals have found only limited specialty uses in wood preservation. Replacement of the current large-scale applications remains a distant goal with these chemistries as their cost-efficacy is disadvantageous in order to achieve the necessary long term performance.

A considerable amount of effort has been put into development of new wood preservatives for various applications. The systems currently under development[26] include organic preservatives such as chlorothalonil and n-octyldichloro-isothiazolinone, waterborne preservatives such as ammoniacal copper based formulations (ACQ, ACB etc.) and light organic solvent preservatives such as tributyltin oxide (TBTO), iodo-propynyl butyl carbamate (IPBC), alkylammonium compounds (AAC), azaconazole and 2-(thiocyano-methylthio)-benzothiazole (TCMTB).

6 BORATES

Borates have been used as industrial wood preservatives for over fifty years, initially in Australia for immunization of susceptible hardwoods from Lyctid attack, and later in New Zealand for the protection of building timbers from wood borers. While interest in borates as low toxicity preservatives has increased sharply in recent years,[27-28] their growth as industrial preservatives outside of the applications described above and anti-sapstain formulations has remained limited to that of an adjunct biocide in preservative formulations where fixation of all the preservative components is unimportant.

When used alone, borates are known to form weak reversible complexes with diols in wood.[29] Similar structures can also form with the borate component during the fixation process of chromated copper borate (CCB) with wood, but these are unfavored and reversible leading to loss of the borate from CCB treatments. Borates may, however, influence the fixation process of the copper and chromium components from CCB in wood differently from that caused to copper and chromium by arsenic in CCA, and so may indirectly improve the performance of copper chromium preservatives. The development of a fixed borate preservative for wood remains an elusive goal, and decreasing the mobility of borate through stronger complexes may lead to a commensurate decrease in biocidal activity caused by the loss of mobility. This is recently being demonstrated in research on complexation of borate with sorbitol.[30]

7 QUATERNARY AMMONIUM COMPOUNDS

Quaternary ammonium compounds (AAC's) have been widely tested as potential wood preservatives.[31] Much of the early research was carried out in New Zealand[32-35] and this led to commercialization of AAC's for the protection of wood used above ground. After problems became apparent with AAC treated wood in service the approval was withdrawn. These problems have been ascribed, in part, to poor distribution of the preservative.[36] Research has sought to explain these problems but no clear solutions have been found.[37-39]

It is believed that quaternary ammonium compounds interact with wood primarily by an ion exchange mechanism.[40] Acidic functionalities in lignin such as carboxylic acids and phenolic hydroxyls may be the reactive sites for such ion exchange. The affinity of quaternary ammonium compounds towards individual wood components follows the order of lignin, hemicellulose and cellulose, and affinity towards lignin increases with increasing pH. The higher affinity of quaternary ammonium compounds towards lignin presumably leads to very low absorption on to cellulose and this may provide little protection against cellulose degradation by brown rot fungi. Recent research[41] has demonstrated the effect of pH on ionization of the protons in acidic groups of the lignin and how this influences the adsorption of quaternary ammonium compounds. At a high pH, quaternary ammonium compounds are totally adsorbed when the concentration in the treating solution is less than the reactive sites available in the lignin. This adsorption of quaternary ammonium compounds is primarily on to carboxylic functionalities as these are much more acidic than phenolic hydroxyl groups. This research showed, however, that only approximately one quat molecule was absorbed for every five C_9 lignin units presumably because of steric effects.

The surfactant nature of quaternary ammonium compounds may have contributed to the higher moisture uptake[42] and partially to the poor weathering performance of AAC treated wood in field tests.[43] Research is in progress to gain a better understanding of these mechanisms and to develop practical solutions.

8 CUPRI-AMMONIUM PRESERVATIVES

Ammoniacal copper preservatives have been examined by Hulme[44] and studied extensively in recent years.[45-51] Copper fixation in cupri-ammonium preservative systems may operate by several different mechanisms. These fixation mechanisms may include the reaction of cupri-ammonium ions with the acidic functional groups in wood by ion exchange, especially with the carboxylic acid groups of lignin and hemicellulose;[52-55] the formation of copper complexes with cellulose through hydrogen bonding between cellulosic hydroxyl group and amine nitrogen;[56-58] through the replacement of one ammonium group in the cupri-ammonium ions with the hydroxyl ion of cellulose;[59] and through the formation of water insoluble copper salts upon evaporation of ammonia.[60]

Studies on ammoniacal copper borate (ACB), ammoniacal copper arsenate (ACA) and ammoniacal copper zinc arsenate (ACZA)[61-63] showed that the co-biocides used in ammoniacal copper systems can influence copper depletion. Similar research with ammoniacal copper quat (ACQ) systems has demonstrated that the copper and quat components influence each other in respect to fixation, distribution and performance.[64-65] In field and fungal cellar tests, the stake moisture content in ACQ treated stakes was found to be influenced by the balance of components in the treatment solution. Furthermore, in leaching studies it was found that the anion portion of the preservative formulation, the cation exchange capacity and the pH of soil were factors which effect leachability. The copper distribution in wood was also influenced by solution pH and the anion moiety.

9. SUMMARY

Understanding the interactions between preservative chemicals and wood components are important to the development of effective wood preservatives and fixation technologies. The development of such knowledge remains a fertile research area, and vital importance as the environmental pressures on the existing preservatives continue to grow.

REFERENCES

1. R.D. Peek and H. Willeitner. 1988. Fundamentals on steam fixation of chromated wood preservatives. Int. Res. Group on Wood Pres. Doc. No.: IRG/WP/3483.
2. H. Willeitner and R.D. Peek. 1988. Less pollution due to technical approaches on accelerated steam fixation of chromated wood preservatives. Int. Res. Group on Wood Pres. Doc. No.:IRG/WP/3483.
3. L. Jin and K.A. Archer. 1991. Copper based wood preservatives: observation on fixation, distribution and performance. Proc. Am. Wood-Preservers' Assoc. 91, (in press).
4. A.R. Zahora and C.M. Rector. 1990. Water repellent additives for pressure treatments. Proc. Can. Wood. Preservation Assoc. 10:1-20.
5. J.A. Butcher and T. Nilsson. 1982. Influence of variable lignin content amongst hardwoods on soft-rot susceptibility and performance of CCA preservatives. Int. Res. Group on Wood Pres. Doc. No:IRG/WP/1151.
6. S.E. Dahlgren. 1975. Kinetics and mechanism of fixation of Cu-Cr-As wood preservatives, part V. Effect of wood species and preservative composition on the leaching during storage. Holzforschung 26(3):84-95.
7. P.M. Rennie, S. Gray and D.J. Dickinson. 1987. Copper based water-borne preservatives: copper adsorption in relation to performance against soft

rot. Int. Res. Group on Wood Pres. Doc. No. IRG/WP/3452.

8. L. Jin and A.F. Preston. 1991 (in press). The interaction of wood preservatives with lignocellulosic substrates. I. Quaternary ammonium compounds. Holzforschung.
9. S.E. Dahlgren and W.H. Hartford. 1972. Kinetics and mechanism of fixation of CCA wood preservatives. I-III Mitt. Holzforschung 26:61-69, 105-113, 142-149.
10. G.B. Fahlstrom, P.E. Gunning and J.A. Carlson. 1967. Copper-Chrome-Arsenate wood preservatives: A study of the influence of composition on leachability. For. Prod. J. 17(7):17-22.
11. W.H. Hartford. 1986. The practical chemistry of CCA in service. Proceedings American Wood Preservers' Association 82:28-43.
12. H. Kubel and A. Pizzi. 1982. The chemistry and kinetic behavior of copper-chromium-arsenic/boron wood preservatives. Part 5. Reactions of a copper-chromium-boron (CCB) preservative with cellulose, lignin and their simple model compounds. Holzforsch. Holzverwert. 34(4):75-83.
13. A. Pizzi. 1990. Chromium interaction in CCA/CCB wood preservatives. Part I. interactions with wood carbohydrates. Holzforschung 44(5):373-380.
14. A. Pizzi. 1990. Chromium interaction in CCA/CCB wood preservatives. Part II. interactions with lignin. Holzforschung 44(6):419-424.
15. A. Pizzi, W.E. Conradie and M. Bariska. 1986. Polyflavonioin tannins - From a cause of CCA soft-rot failure to the "missing link" between lignin and microdistribution theories. Int. Res. Group on Wood Pres. Doc. No.: IRG/WP/3359.
16. D.V. Plackett. 1983. A discussion of current theories concerning CCA fixation. Int. Res. Group on Wood Pres. Doc. No. IRG/WP/3238.
17. K. Yamamoto and M. Inoue. 1990. Difference of CCA efficacy among coniferous wood species. Int. Res. Group on Wood Pres. Doc. No. IRG/WP/3601.
18. M.A. Hulme and J.A. Butcher. 1977. Soft rot control in hardwoods treated with chromated copper arsenate preservatives. III. Influence of wood substrate and copper loadings. Mater. Org. 12:223-234.
19. H. Greaves. 1977. An illustrated comment on the soft rot problem in Australia and Papua New Guinea. Holzforschung 31:71-79.
20. H. Greaves. 1972. Structural distribution of chemical components in preservative-treated wood by energy dispersion X-ray analysis. Mater. Org.

7:277-286.
21. H. Greaves. 1974. The microdistribution of copper-chrome-arsenic in preservative treated wood using X-ray microanalysis in scanning electron microscopy. Holzforschung 27:80-88.
22. D.J. Dickinson. 1974. The microdistribution of copper-chrome-arsenate in Acer pseudoplatanus and Eucalyptus maculata. Mater. Org. 9:21-33.
23. T. Nilsson. 1982. Comments on soft rot attack in timbers treated with CCA preservatives: A document for discussion. Int.Res. Group on Wood Pres., Doc.No. IRG/WP/1167.
24. G. Daniel and T. Nilsson. 1987. Comparative studies on the distribution of lignin and CCA elements in birch using electron microscopic X-ray microanalysis. Int. Res. Group on Wood Pres. Doc. No:IRG/WP/1328.
25. A.F. Preston. 1983. Dialkyldimethylammonium Halides as Wood Preservatives. J. Am. Oil Chem. Soc. 60(3):567-570.
26. A.F. Preston. 1987. New protection agents for wood products. FPRS Proc.47358 Conference on "Wood protection techniques and the use of treated wood in construction" pp.42-47.
27. D.D. Nicholas, L. Jin and A.F. Preston. 1990. Immediate research needs for diffusible boron preservatives. FPRS Proc. 47355. First international conference on wood protection with diffusible preservatives. pp. 121-123.
28. J.D. Lloyd and D.J. Dickinson. 1991. Comparison of the inhibitory effects of borate, germanate, tellurate, arsenite and arsenate on 6-phosphogluconate dehydrogenase. Int.Res. Group on Wood Pres., Doc.No. IRG/WP/1508.
29. J.D. Lloyd, D.J. Dickinson and R.J. Murphy. 1990. The probable mechanism of action of boric acid and borates as wood preservatives. Int.Res. Group on Wood Pres., Doc.No. IRG/WP/1450.
30. J.D. Lloyd, D.J. Dickinson and R.J. Murphy. 1991. The effect of sorbitol on the decay of boric acid treated scots pine. Int.Res. Group on Wood Pres., Doc.No. IRG/WP/1509.
31. H. Becker. 1989. Alkylammoniumverbindungen als holzschutzmittel. Seifen-Oele-Fette-Wachse 115(18):681-684.
32. J.A. Butcher and J.A. Drysdale. 1977. Relative Tolerance of Seven Wood-Destroying Basidiomycetes to Quaternary Ammonium Compounds and Copper-Chrome-Arsenate Preservative. Mater. Org.

12(4):271-277.
33. J.A. Butcher and J. Drysdale. 1978. Efficacy of acidic and alkaline solutions of alkylammonium compounds as wood preservatives. New Zealand J. For. Sci. 8(3):403-409.
34. J.A. Butcher, A. F. Preston and J.A. Drysdale. 1979. Potential of unmodified and copper-modified alkylammonium compounds as groundline preservatives. New Zealand J. For. Sci. 9(3): 348-358.
35. J.A. Butcher and H.Greaves. 1982. AAC Preservatives: Recent New Zealand and Australian Experience. Int. Res. Group on Wood Pres., Doc.No. IRG/WP/3188.
36. J.A. Butcher. 1985. Benzalkonium chloride (an AAC preservative); criteria for approval, performance in service, and implications for the future. Int.Res.Group on Wood Pres. Doc.No. IRG/WP/3328.
37. J.N.R. Ruddick. 1984. The Influence of Staining Fungi on the Decay Resistance of Wood Treated with Alkylammonium compounds. Mater. Org. 21(2):139-149.
38. J.N.R. Ruddick and A.R.H. Sam. 1982. Didecyldimethylammonium chloride-a quaternary ammonium wood preservatives: its leachability and distribution in, four softwoods. Mater. Org. 17(4):299-313.
39. D.D. Nicholas, A.D. Williams, A.F. Preston and S. Zhang. 1990. Distribution and Permanency of Didecyldimethylammonium chloride in Southern Pine Sapwood Treated by the Full-Cell Process. For.Prod.J. 41(1):41-45.
40. A.F. Preston, P.J. Walcheski, P.A. McKaig and D.D. Nicholas. 1987. Recent Research on Alkylammonium Compounds in the U.S. Proc. Am. Wood Pres. Assn. 83:331-348.
41. L. Jin and A.F. Preston. 1991 (in press). The interaction of wood preservatives with lignocellulosic substrates. I. Quaternary ammonium compounds. Holzforschung.
42. L. Jin and K.A. Archer. 1991. Copper based wood preservatives: observation on fixation, distribution and performance. Proc. Am. Wood-Preservers' Assoc. 91, (in press).
43. L. Jin, K.A. Archer and A.F. Preston. 1991. Surface characteristics of wood treated with various AAC, ACQ and CCA formulations after weathering. Int. Res. Group on Wood Pres. Doc. No:IRG/WP/2369.
44. M.A. Hulme. 1979. Ammoniacal wood preservatives. Rec. Ann. Conv. Brit. Wood-Pres' Assoc. 38-50.
45. H. Greaves, N. Adams and D.F. McCarthy. 1982. Studies of preservative treatments for hardwoods in

ground contact. Holzforschung 36:225-231.

46. B. Henningsson, B. Hager and T. Nilsson. 1980. Studies on the protective effect of water-borne ammoniacal preservatives on hardwoods in ground contact situations. Holz als Roh-und-Werkstoft 38: 95-100.

47. A.F. Preston, P.A. McKaig and P.J. Walcheski. 1985. Recent studies with ammoniacal copper carboxylate preservatives. Proc. Am. Wood-Preservers' Assoc. 81: 30-39.

48. B.R. Johnson and D. Gutzmer. 1978. Ammoniacal copper borate: a new treatment for wood preservation. For. Prod. J. 28(2):33-36.

49. B.R. Johnson. 1983. Field trials with ammoniacal copper borate wood preservative. For. Prod. J. 33(90):59-63.

50. J. Rak and H. Unligil. 1978. Fungicidal efficacy of ammoniacal copper and zinc arsenic preservatives tested by soil block cultures. Wood Fiber 9: 270-275.

51. L.R. Wallace. 1986. Mathematical modelling in fungus cellar screening of ammoniacal Copper/Quaternary ammonium preservative systems efficacy. Proc. Am. Wood-Preservers' Assoc. 82: 159-171.

52. J.A. Butcher and T. Nilsson. 1982. Influence of variable lignin content amongst hardwoods on soft-rot susceptibility and performance of CCA preservatives. Int. Res. Group on Wood Pres. Doc. No:IRG/WP/1151.

53. L. Jin, D.D. Nicholas and T.P. Schultz. 1990. Dimensional stabilization and decay resistance of wood treated with brown-rotted lignin and copper sulfate. Int. Res. Group on Wood Pres. Doc. No:IRG/WP/3608.

54. S.A. Boyd, L.E. Sommers and D.W. Nelson. 1981. Copper (II) and iron (III) complexation by the carboxylate group of humic acid. Soil Sci. Soc. Am. J. 45:1241-1243.

55. A. Myers and R.D. Preston. 1961. Sorption of copper by cellulose. Nature 190:803-804.

56. O. Hinojosa, J.C. Arthur and T. Mares. 1974. Electron spin resonance studies of interactions of ammonia, copper, and cupriammonia with cellulose. J. Appl. Poly. Sci. 18:2509-2516.

57. W.E. Davis, A.J. Barry, F.C. Peterson and A.J. King. 1943. X-ray studies of reactions of cellulose in non-aqueous systems. II. Interaction of cellulose and primary amines. J. Am. Chem. Soc. 65:1294-1299.

58. H.B. Jonassen and T.H. Dexter. 1949. Inorganic complex compounds containing polydentate groups. I. The complex ions formed between copper (II) ions and ethylenediamine. J. Am. Chem. Soc. 71:1553-1571.
59. P.J. Baugh, O. Hinojosa, J.C. Arthur and T. Mares. 1968. ESR spectra of copper complexes of cellulose. J. Appl. Poly. Sci. 12:249-265.
60. M.A. Hulme. 1979. Ammoniacal wood preservatives. Rec. Ann. Conv. Brit. Wood-Pres' Assoc. 38-50.
61. B.R. Johnson and D. Gutzmer. 1978. Ammoniacal copper borate: a new treatment for wood preservation. For. Prod. J. 28(2):33-36.
62. B.R. Johnson. 1983. Field trials with ammoniacal copper borate wood preservative. For. Prod. J. 33(90):59-63.
63. C.W. Best and G.D. Coleman. 1981. AWPA Standard M11: an example of its use. Proc. Am. Wood Preservers' Assc. 81:35-42.
64. L. Jin and K.A. Archer. 1991. Copper based wood preservatives: observation on fixation, distribution and performance. Proc. Am. Wood-Preservers' Assoc. 91, (in press).
65. L. Jin and A.F. Preston. 1991 (in press). The interaction of wood preservatives with lignocellulosic substrates. I. Quaternary ammonium compounds. Holzforschung.

Waterbased Fixed Preservatives

D.G. Anderson, J.A. Cornfield, and G.R. Williams

HICKSON TIMBER PRODUCTS LTD., WHELDON ROAD, CASTLEFORD WF10 2JT, UK

INTRODUCTION

Most waterbased fixed preservatives in commercial use are based on the readily available, and cost effective biocide copper. The world market for industrial wood preservatives is dominated by the group of preservatives known generally as 'chromated copper'. Amongst this group are chromated copper arsenate (CCA), chromated copper borate (CBC) and chromated copper silicafluoride (CFK). A second less extensive group, but still of commercial significance is that based on ammoniacal copper.

In the 1920's, chromated copper was patented by Gunn. This was the first commercially recorded use of chromated copper as a system for chemical fixation of copper. In the 1930's a major breakthrough occurred when Kamesan Sonti patented chromated copper arsenate (CCA). From its modest origins in India, CCA has become the dominant fixed copper preservative. Sales world wide in 1987 are estimated to have exceeded £150,000,000, and in excess of 20 million m^3 of wood was treated. This quantity of treated wood could have formed an 80ft wide path of one inch thick material around the equator.

The success of such a simple single product will be explained by some of the following sections relating to the key properties of these products.

Dissolution of copper salts in ammoniacal liquor has been used as an alternative means of solubilising some of the more insoluble copper salts for example arsenate, borate and carbonate, to allow them to be used for wood preservation. These systems are less widely used but provide an alternative physical method for achieving copper fixation. The most widely used are ammoniacal copper arsenate (ACA) and ammoniacal copper zinc arsenate (ACZA). These products are most used in U.S.A and Canada.

Ammoniacal copper caprylate has been used in Scandinavia and more recently ammoniacal copper carbonate with quaternary ammonium compound as co-biocide is being promoted.

The commercial success of waterbased wood preservatives is based on a number of key features. These are:

- solubility in treatment solutions
- insolubility in treated timber
- low volatility of active ingredients
- fungicidal efficacy
- insecticidal efficacy
- low corrosivity
- ready availability
- low cost
- acceptable toxicity/bio availability

Before discussing these features and their interactions with wood, it is important to understand the structure of the softwood substrate, particularly in relation to the chemical interactions of wood preservative components and biochemical interactions of the fungi degrading the substrates.

SOFTWOOD STRUCTURE AND CHEMISTRY

The softwood cell wall accounts for 98% of the wood material. The cell walls are layered structures composed of polymeric compounds, primarily cellulose fibrils interspersed with hemi-celluloses and lignin. They surround the lumen. The lumen form the main channels for flow of sap or preservative solution through the timber, and are the main pathways by which fungi can colonise the wood structure. Flow of materials between adjacent cells occurs via openings in the cell wall known as pits. Between the cells is the middle lamella. This is amorphous and contains a high concentration of lignin. A typical chemical analysis for Scots Pine is 52% cellulose, 22% hemicellulose, 26% lignin.

Linear cellulose chains are aligned and stiffened by the formation of intra and inter molecular hydrogen bonds to form the fibrils which make up the cell wall layers. Preservative penetration into the cell wall is assisted if it is applied from a swelling solvent such as water. The solvent is adsorbed via hydrogen bonds onto the cellulose, especially in the least crystalline zones and so swells the cell wall forming transient capillaries. Other small hydrogen bond forming molecules such as ammonia can also act to swell the cell wall and assist in micro distribution of the preservative.

Hemicelluloses are polysaccharides composed of various sugar units, with shorter molecular chains than cellulose and some branching of the chains. The hemicelluloses do not pack as well as cellulose polymers, and they are more available than cellulose both for decay by fungi and for interaction with preservative chemicals. The principle sugars found in Scots Pine beside glucose are mannose 12.4%, xylose 7.6%, and glucuronic acid 5%.

Lignin is a polyphenolic branched polymer which fills the spaces between the polysaccharide fibrillar elements of the cell walls. Coniferyl alcohol is the predominant precursor of softwood lignin. During production of lignin this reacts by an enzymatic dehydrogenation to form phenoxyradicals, which couple randomly to form a large amorphous, three dimensional polymer.

Typical functional groups on lignin are phenolic and aliphatic hydroxyl groups, methoxyl groups, carbonyl and alkene groups.[1] Both existing functional groups and new groups formed by reaction of lignin with reactive preservatives play an important part in wood - preservative interactions.

DECAY OF TIMBER BY FUNGI

Polysaccharides and lignin are degraded by different biochemical reaction mechanisms according to the type of fungi which are present in the substrate. Each type will have a different effect on the various structural components in accordance with their chemical composition and availability. Degrade by different fungi, therefore, results in a different visual appearance of the substrate and it is by this means that they are classified into the three main groups; brown rot, white rot and soft rot.

For the timber to be effectively protected, the preservative active ingredients must be located at sites within the timber which the fungi are able to attack. The degrade of timber by brown rot fungi (basidiomycotina) is characterised by breakdown of the polysaccharide portions of the cell wall leaving lignin relatively unaffected. Initial degradation would appear to be within the secondary cell wall resulting initially from the action of an oxidative process involving hydrogen peroxide and ferrous ions which readily diffuse into the cell wall, and thereafter by cellulase enzyme degradation.[2] The white rot fungi (basidiomycotina) have the ability to utilise both cellulose and lignin components. Initial degrade involves primarily the lignin components and is caused by oxidative processes involving opening of the aromatic ring either before or after breakdown into simpler molecules. Lignin peroxidases and metal dependant extra-cellular enzymes which generate hydrogen peroxide are instrumental in causing these reactions to occur.[3] Rapid lignin depolymerisation in many cases results in middle-lamella degradation and hence the fibrous nature of the wood after attack by these fungi. In contrast, the soft rot fungi (ascomycotina, deuteromycotina) have a more distinct mode of attack forming cavities within the secondary cell wall where they principally degrade polysaccharide materials although some have been shown to have weak lignin degrading systems. The ability of soft rot fungi to grow within the cell wall has important implications in terms of the distribution of wood preservative components.

The interaction of wood preservatives and particularly copper and fungi, is important for understanding mechanisms of activity. At the simplest level, the presence of copper can act selectively by inhibiting other timber degrading fungi and selecting for copper tolerant species. This disturbs the natural successional pattern of colonisation and allows these tolerant organisms to grow more rapidly in the absence of competing 'non tolerant' organisms.

Copper acts as a fungicide by precipitation of proteins within the fungi and by interference with enzyme reactions. Certain fungi, particularly brown rots and soft rots are relatively tolerant to the presence of copper and higher loadings are required to produce a fungicidal effect. The brown rot Poria placenta has been shown to precipitate copper in an insoluble form as the oxalate which effectively prevents copper from interfering with fungal metabolism.[4] Soft rot fungi have been shown to have similar mechanisms but involve the formation of copper complexes with compounds such as polyphosphate which again removes the fungicidal effectiveness of the copper. Such is the ability of the micro organisms to cope with relatively high levels of metals (up to 5.0 kgm^{-3} of copper), that a secondary biocide may be required to achieve an adequate spectrum of activity. This is discussed more fully below.

PRESERVATIVE REQUIREMENTS

From this consideration of timber structure and modes of action of wood decay fungi and tolerance to copper, the requirements for use of copper as an effective fixed water based preservative can be derived.

1. An additional fungicide should be present to protect from attack by copper tolerant fungi.

2. Copper should be well distributed throughout the timber structure to prevent internal decay of large timber sections.

3. Penetration of the preservative into the cell wall may be required to protect against decay.

4. The copper should be well fixed to ensure permanent protection under the leaching conditions which occur in use in or out of ground contact.

The ways in which each of these requirements are met by commercially available water based fixed preservatives is discussed in the following sections.

TYPE OF COPPER CONTAINING PRESERVATIVES AND SELECTION OF ADDITIONAL PRESERVATIVE ELEMENTS

Two types of aqueous preservative whose efficacy against fungi is based predominantly on the presence of copper are available. The most commonly used types are acidic formulations of dissolved copper salts containing chromium which serves to prevent corrosion of mild steel by the formulations and to fix the copper permanently in the timber. Arsenic is often added to protect against copper tolerant fungi and insects to give copper, chromium, arsenic preservatives (CCA).

Formulations of this type are highly acidic in character pH 1.5-3. Other examples are copper chromium, copper chromium boron, copper chromium fluorine and copper chromium phosphorus. Of these formulations, CCA is the most effective due to the superior efficacy and permanence of the arsenic component.

Another type of formulation not containing chromium is based on the ability of ammonia to complex the copper and maintain it in solution at a high pH. Arsenic has again been found effective as an additional biocide giving the commercially available formulations ammoniacal copper arsenate ACA, and ammoniacal copper zinc arsenate ACZA. Other ammoniacal formulations which have been used commercially include ammoniacal copper caprylate and ammoniacal copper carbonate and quaternary ammonium compounds.

DISTRIBUTION OF COMPONENTS

Treatability of timber and distribution of preservative depends on the timber structure. Structure of hardwoods varies from that described for softwoods, resulting in considerable difference in treatability. Due to the greater commercial significance of softwood treatment, distribution of preservatives in softwoods only is considered here.

Good liquid flow in permeable softwoods ensures good penetration of CCA preservatives throughout the sapwood.[5]

Micro distribution of CCA has also been extensively studied. Petty and Preston,[6] showed that copper cannot penetrate the crystalline regions of cellulose fibrils but can penetrate less regular regions of cellulose and hemicellulose. Chou, Chandler and Preston,[7] showed that copper was present in the cell wall and that the lumen surface was well protected. Greaves,[8] showed pit membranes also contained high levels of copper. Copper is, therefore, well distributed throughout the macro and micro structure of CCA treated Scots Pine.

Much of the work done on ammoniacal copper containing preservatives has used refractory species such as Spruce. Rak and Clark,[9] recorded improvements in penetration over that obtained with CCA. However, penetrability of Scots Pine with ammoniacal preservatives is inferior to that with CCA resulting in untreated sapwood at the centre of timber sections.[10] The reason for this has been hypothesised to be due to the capacity for ammonia to swell polysaccharide regions within the cell resulting in blocking of the pits and prevention of free liquid flow.

Rak,[11] has reported work on the solvating power of anhydrous liquid ammonia which swells amorphous cellulose, hemicellulose, lignin and crystalline cellulose. He showed that even with the low concentrations of aqueous ammonia used to treat Spruce, micro structural damage could follow treatment allowing penetration of copper into the cell walls and intercellular regions.

Thus with permeable species such as Scots Pine, micro distribution in the treated areas is good for both ammoniacal and acidic copper based preservatives. However, CCA offers better protection from decay due to the superior penetration throughout the sapwood.

COPPER PERMANENCE

CCA preservatives are available with various chemical compositions. Table 1 shows the percentage composition of three types approved by the American Wood Preservers Association.

Table 1 CCA Oxide formulations approved by the AWPA, percentage composition

	Type A	**Type B**	**Type C**
CrO_3	65.5	35.3	47.5
CuO	18.1	19.6	18.5
As_2O_5	16.4	45.1	34.0

In the U.K. preservatives are usually prepared from sodium dichromate, copper sulphate and arsenic pentoxide using a similar balance of copper:chromium:arsenic as listed above for AWPA Type C.

Much work has been done to determine the permanence of copper, chromium and arsenic when applied from various CCA formulations under conditions of leaching with water.[12-20] These studies have generally shown that formulations of the following composition corresponding to CCA Salts Type 2 have the greatest permanence.

Sodium dichromate. $2H_2O$	45%
Copper sulphate. $5H_2O$	35%
Arsenic pentoxide. $2H_2O$	20%

All the elements are essentially fixed with around two percent of the copper removable by cold water leaching under neutral conditions.

Copper permanence after application from ammoniacal solutions has also been studied.[9,21-23] Results vary but in general the level of copper permanence was similar to that found for CCA where comparisons were made.

FIXATION OF COPPER FROM AMMONIACAL SYSTEMS

The mechanism by which the elements are rendered permanent in the timber is generally referred to as 'fixation'. Whilst fixation mechanisms for CCA have been studied extensively, relatively little work has been done on mechanisms of fixation from ammoniacal systems. This is believed to involve evaporation of all or part of the ammonia and precipitation of insoluble copper hydroxide, or copper arsenate in ammoniacal copper arsenate. In addition, part of the copper fixation may be through ion exchange at hydroxy or carboxylic anionic sites on the timber which will tend to be ionised under the high pH conditions generated.

FIXATION OF COPPER FROM CHROMATED SYSTEMS

Due to the commercial importance of CCA and its remarkable properties of permanence several studies have been done to elucidate the fixation mechanism. Different methods have been employed by various workers. Dahlgren,[24-29] measured change in pH on reaction of CCA with wood flour, he also tested leachability of chromium during the course of fixation to determine rate constants for the various reactions. Pizzi,[30-33] analysed insoluble reaction products formed during the reaction of CCA with simple model compounds selected to represent the building blocks of cellulose and lignin. In more recent work Hon and Chang,[34] studied solution interactions of model compounds with chromium spectroscopically. Spectroscopic techniques have also been used by Ostmeyer et al,[35,36] and Plackett, [37] in an attempt to characterise reaction products of CCA in solid wood. Pizzi,[38,39] has calculated lowest energy states of interactions between chromium and carbohydrates and lignin model compounds to further elucidate the mechanisms of reaction in CCA wood preservatives. The rate of fixation of CCA is strongly temperature dependant, and this and other factors affecting fixation have been reviewed by Anderson.[40]

CCA Reaction Rate Studies

Dahlgren,[24-29] tested leachability of chromium during the course of fixation of CCA in sawdust. Rate constants for fixation were found to vary during the course of the reaction and formation of intermediates and conversion to final products was proposed. pH measurements showed an immediate increase due to ion exchange and adsorption reactions with the wood. When the pH reached a maximum good fixation of copper, chromium and arsenic had occurred.

Table 2 Summary of the proposed species present during the fixation of CCA, from work by Dahlgren

PRECIPITATION PERIOD

pH	Time	Species	Formula
2.5-3	0-30 min	Complex chrome chromates	$[Cr(III)_2(Cr(VI)O_4)_3]$ --acidic wood groups
		Acid copper chrome arsenate	$HCuCr(III)(AsO_4)$ Variable ratios
		Acid copper arsenate	$HCu(AsO_4)$ Variable ratios
3-4	1-12 hr	Tertiary copper chrome arsenate	$CuCr(III)(AsO_4)$ Variable ratios
4-5	12-100hr	Copper arsenate	$Cu_3(AsO_4)_2$
		Chrome arsenate	$Cr(III)AsO_4$
		Basic copper arsenate	$Cu(OH)CuAsO_4$
5-5.5	100-250hr	Basic chrome chromates	$Cr(III)_4(OH)_{10}Cr(VI)O_4$ $[Cr(III)(OH)_2]_2Cr(VI)O_4$

FINAL CONVERSION PERIOD

Acid and tertiary copper arsenate ⟶ basic copper arsenate

Chromates + wood complexed chromates ⟶ tertiary chrome $Cr(OH)_3$

FINAL EQUILIBRIUM PRODUCTS :

Cu fixed by ion exchange
Chrome arsenate $Cr(III)AsO_4$
Basic copper arsenate $Cu(OH)CuAsO_4$
Chrome hydroxide $Cr(III)(OH)_3$

CCA solutions were also reduced with hydrogen peroxide at various pH levels obtained by partial neutralisation with caustic soda. No timber or wood sugars were present during these experiments. Analysis of the precipitates showed that at low pH an acid arsenate containing chromium and some copper was formed, at higher pH, monohydrogen copper arsenate precipitated.

A complex reaction scheme was hypothesised by Dahlgren which identified three key stages in the reaction; adsorption, reduction, and final fixation. A summary of the proposed species present at different times after the commencement of the reaction and the pH at each stage is given in Table 2.

In this work the nature of the bonding to wood was not investigated.

CCA Reactions with Model Compounds

Pizzi,[30-33] used a mixed salt/oxide preservative $CuSO_4 5H_2O$, CrO_3, As_2O_5 with similar Cu:Cr:As molar ratios to those used in CCA Salts Type 2 ie. 1:2:1.1. Simple model compounds guaiacol and D(+) glucose were used to represent the lignin and cellulose components of wood, wood flour was also used.

The rate of chromium (VI) reduction to chromium (III) was followed spectrophotometrically and rate constants for the various reactions calculated.

When copper, chromium, arsenic and model compounds or mixtures were present in solution the pH increased and precipitates formed. These were analysed and the compositions and postulated major components are given in Table 3.

Table 3 Composition of precipitates on interaction of CCA with model compounds, found by Pizzi

	% w/w in precipitate						
Model Compound	Cu	Cr	As	C	H	O	Major Component
D(+) glucose	1	21	23	3	2	49	Inorganic chromium arsenate
Guaiacol	3	16	15	24	3	40	Chromium arsenate guaiacol complex
Mixture	2	20	20	15	3	40	

From these results and associated studies, it was predicted that after fixation of CCA Type C all the arsenic and hence fifty percent of the chromium was present as chromium arsenates Cr(III) AsO_4. These were partly complexed with the lignin and partly precipitated onto the cellulose as inorganic salts. Ninety percent of the copper was said to be bound to the cellulose and the lignin. Ten percent of the copper was thought to be present as copper chromate bound to the lignin CuCr(VI)O_4 with the remaining chromium being present as inorganic precipitates of chromium containing complexes on cellulose.

The work of Hon and Chang,[34] using electron spin resonance spectroscopy (ESR) confirmed that chromium glucose complexes do not form in solution. However when cellobiose, the disaccharide from which cellulose is formed, was tested in the presence of chromium in solution, weak complexes of chromium and cellobiose did form in solution.

Calculations by Pizzi,[38] have shown that a transition complex of chromium (VI) with disaccharide units is energetically favoured. He postulates that chromium reduction occurs at these sites to form mixed Chromium (VI) Chromium (III) disaccharide transition complexes. In the presence of arsenate ions the Chromium (III) species are abstracted to form chromium arsenates until nearly all the Chromium (VI) has been reduced. Calculations show that chromium (III) arsenates are only weakly adsorbed on carbohydrates in agreement with previous findings.

Further calculations on interactions of guaiacol with chromium VI by Pizzi,[39] have shown that a favoured energy state is possible giving support to the possibility of reaction of CCA with lignin.

<u>Reaction Products of CCA with Solid Wood</u>

Work by Ostmeyer and Winandy[35,36] has addressed the question of determining the effect of the preservative elements on the timber components. They used X-ray photo electron spectroscopy (XPS) and diffuse reflectance fourier transform infra red spectroscopy (DRIFT) to examine solid treated timber.

XPS gives information about oxidation in timber, in particular, whether carbon atoms are bonded to carbon and hydrogen, or hydroxyl, or carbonyl oxygens. Due to the existence of all these types of carbon bonding in both the cellulosic and lignaceous regions, the spectrum is composed of over-lapping signals. Timber treated with CCA showed an increase in the number of carbon atoms bonded to hydroxyl groups. Oxidation of hydroxy groups to carboxyl groups was not detected.

DRIFT gave inconsistent spectra for tree growth produced in the summer. Results from growth produced in the spring showed a decrease in aromatic and carbonyl content, presumed to be by bond formation at these sites.

These findings are consistent with oxidation of aromatic and unsaturated areas such as those occurring in the lignin structure. They do not suggest extensive oxidation in the carbohydrate areas which would be associated with an increase in carbonyl content.

Plackett,[37] used electron spin resonance spectroscopy (ESR) to give information on the molecular environment of atoms with unpaired electrons, such as copper (II) in timber. He found that after treatment of timber with copper sulphate a similar spectrum was found to when cellulose was treated with copper sulphate. Treatment of lignin gave a different spectrum. Results for CCA treated sapwood showed that hydrated copper was present as with copper sulphate treatments, but no evidence for copper location when applied from CCA solution was found in the work reported.

CCA Reaction Mechanisms

From the work of Dahlgren it is evident that reduction of chromium (VI) to chromium (III) is the essential reaction occurring during CCA fixation. The redox reaction consumes hydrogen ions and causes the pH to increase inside the timber during fixation.

Both the work of Dahlgren and that of Pizzi showed that chromium arsenate was one of the main reaction products. It was found by Pizzi that the chromium (VI) could be reduced both in the presence of cellulosic and lignin like model compounds, but that the chromium arsenate formed interacted more strongly with lignin like compounds. The work of Hon and Chang lends support to the theory of weak chromium cellulose interactions. Thus in the carbohydrate areas it appears likely that metal arsenates are mainly inorganic in character with relatively weak association with the organic polymer. Organic-inorganic interactions may, however, be stronger for lignin.

The principle fungicidal element copper is present in more than one form, the balance between the various forms depending on the exact CCA composition and timber species. From the work of both Dahlgren and Pizzi the fixation mechanism for a substantial proportion of the copper was proposed to be by ion exchange. Simple copper salts such as copper sulphate are also believed to undergo ion exchange in the timber, but these simple salts are less tightly bound. The reason for the superior fixation of copper through ion exchange, when applied from CCA solution, is postulated to be due to an increase in the number of ion exchange sites and their affinity for copper caused by oxidation of the timber components during reduction of chromium (VI). Formation of insoluble copper arsenate will also contribute to copper fixation.

Ostmeyer and Winandy showed that after reaction of the timber with CCA there was a significant increase in the number of hydroxyl groups in the timber. Considering the chemical composition of the wood components it is evident that oxidative reactions in the cellulose or hemicellulose would lead to an increase in carboxyl content and would not explain the findings. Therefore it appears unlikely that carbohydrate regions are extensively and permanently oxidised during CCA fixation. The increase in hydroxy groups detected is more consistent with oxidation in the lignin, which may serve to provide stronger affinity for copper ions.

Amongst the authors there is a consensus that the following factors are most important :

- The redox reaction between Cr(VI) and wood consumes H^+ thereby increases pH and is the essential component of fixation since it causes the inorganic metallic arsenates to become insoluble.

- Inorganic metal arsenates precipitation forms the largest proportion of reaction products.

- The chemical components of wood are modified by CCA particularly in the lignin regions thus enabling copper to be more effectively bound to wood.

- Wood/metal/arsenate complexes assist in binding the metals more effectively.

BIOLOGICAL IMPLICATIONS

Copper based formulations can suffer from reduced performance against copper tolerant organisms, or may fail due to poor penetration.

It is interesting to note that many of the copper tolerant fungi are brown rots such as Coniophera puteana and Poria placenta which principally degrade cellulose and hemi-cellulose. White rot fungi which can degrade lignin are not usually resistant to copper in treated softwoods. This may lend support to the possibility that lignin plays an important part in the reactions of CCA in softwood.

The role of arsenic in CCA is often thought of only as an insecticide or fixing agent, neglecting its very important role as a secondary biocide, particularly effective against many of the copper tolerant organisms.

The search for copper based alternatives to CCA needs, therefore, to recognise the extremely valuable and cost effective role of arsenic in CCA and to make accordingly substantial efforts to extend the range of efficacy of the formulations.

It has also been found that many alternative organic biocides are more susceptible to biodetoxification than CCA. This also needs to be taken account of in researching new timber preservatives.[41]

CONCLUSIONS

From the above review of studies aimed at identifying the reactions and interactions arising from treatment of wood with both ammoniacal copper and chromated acidic copper the main factors affecting performance are considered to be:

a) Penetration of the preservative into the wood structure.

b) Broadening of spectrum of activity of the preservatives to cope with copper tolerant organisms.

c) Fixation of the fungicidal elements.

The main areas still not clearly understood are the chemical environment of copper in fixed preservatives, and the role fixation agents play either directly or indirectly in influencing the performance of the copper in the preservative.

The preservative requirements derived from a consideration of the nature of timber and mechanisms of decay are currently well satisfied by CCA. Arsenic is present both to protect timber from decay by copper tolerant fungi and to act as an insecticide. Copper is well distributed throughout the macro and micro structure of the timber, including penetration into the cell wall. The copper and arsenic are well fixed in CCA, particularly when the copper, chromium and arsenic components are present in the same ratios as CCA Salts Type 2.

REFERENCES

1. Q. Fengel and G. Wegener, 'Wood Chemistry, Ultrastructure, Reactions', Walter de Gruyter, 1989
2. W. Liese and R. Schmid, Phytopathologische Zeitschrift, 1964, 51, 385
3. P. Bonnarme and T.W. Jeffries, App. Env. Microbiol., 1990, 210
4. H.P. Sutter and G.E.B. Jones, Rec. Ann. Conv. Br. Wood Preserv. Assoc., 1985, 3, 29
5. HTP Internal Report, P.S. Warburton, (unpublished), 1988
6. J.A. Petty and R.D. Preston, Holzforschung, 1968, 22, 174
7. C.K. Chou, J.A. Chandler and R.D. Preston, Wood Sci. Technol., 1973, 7, 151
8. H. Greaves, Holzforschung, 1974, 28, 193
9. J.R. Rak and M.R. Clark, Ann. Meet. Am. Wood Preserv. Assoc., 1974, 27
10. HTP Internal Report, A. Lyman, (unpublished), 1988
11. J.R. Rak, Holzforschung, 1977, 3, 29
12. G.D. Fhalstrom, For. Prod. J., 1967, 7, 17
13. W.T. Henry and E.B. Jeroski, Annu. Meet. Am. Wood-Preserv. Assoc., 1967, 63, 187
14. B. Hager, For. Prod. J., 1969, 19, 21
15. E.M. Wallace, Annu. Meet. Am. Wood-Preserv. Assoc., 1968, 50
16. D.N.R. Smith and A.I. Williams, Wood Sci. Technol., 1973, 7, 60
17. D.N.R. Smith and A.I. Williams, Wood Sci Technol., 1973, 7, 142
18. W.S. McNamara, Int.Res. Group Wood Preserv. Ann. Conf. IRG 20, 1989, IRG/WP/3054
19. W.S. McNamara, Int. Res. Group Wood Preserv. Ann. Conf. IRG 20, 1989, IRG/WP/3505
20. HTP Internal Report, P.S. Warburton, (unpublished), 1988
21. M.A. Hulme, Rec. Ann. Conv. Br. Wood Preserv. Assoc., 1979, 38
22. C.E. Sudman, Int. Res. Group Wood Preserv. Ann. Conf.

IRG 15, 1984, IRG/WP/3299
23. P. Vasishth, R.C. Vasishth and D. Nicholas, Proc. Conf. Wood Protection with Diffusible Preserv., 1990, 35
24. S.E. Dahlgren, Rec. Ann. Conv. Br. Wood Preserv. Assoc., 1972, 8, 109
25. S.E. Dahlgren, W.H. Hartford, Holzforschung, 1972, 26, 62
26. S.E. Dahlgren, W.H. Hartford, Holzforschung, 1972, 26, 105
27. S.E. Dahlgren, W.H. Hartford, Holzforschung, 1972, 26, 142
28. S.E. Dahlgren, Holzforschung, 1974, 28, 58
29. S.E. Dahlgren, Holzforschung, 1975, 29, 84
30. A. Pizzi, J. Polym. Sci. Polm. Chem. Ed., 1981, 19, 3093
31 A. Pizzi, J. Polym. Sci. Polm. Chem. Ed., 1982, 20, 707
32. A. Pizzi, J. Polym. Sci. Polm. Chem. Ed., 1982, 20, 725
33. A. Pizzi, J. Polym. Sci. Polym. Chem. Ed., 1982, 20, 739
34. D.N.S. Hon and S.T. Chang, Wood Fiber Sci., 1985, 17, 92
35. J.G. Ostmeyer, T.J. Elder, D.M. Littrell, B.J. Tatarchuck, J.E. Winandy, J. Wood Chem. Technol., 1988, 8, 413
36. J.G. Ostmeyer, T.J. Elder, J.E. Winandy, J. Wood Chem. Technol., 1989, 9, 105
37. D.V. Plackett, E.W. Ainscough and A.M. Brodie, Int. Res. Group Wood Preserv. Ann. Conf. IRG 18, 1987, IRG/WP/3423
38. A. Pizzi, Holzforschung, 1990, 44, 373
39. A. Pizzi, Holzforschung, 1990, 44, 419
40. D.G. Anderson, Proc. Can. Wood Preserv. Assoc., 1989, 75
41. P.A. Briscoe, G.R. Williams, D.G. Anderson and G.M. Gadd, Int. Res. Group Wood Preserv. Ann. Conf. IRG 21, 1990, IRG/WP/1464

Wood Preservation: Strategies for the Future

P. Vinden and J.A. Butcher

WOOD TECHNOLOGY DIVISION, FOREST RESEARCH INSTITUTE, ROTORUA, NEW ZEALAND

1 ABSTRACT

The preservative treatment of wood has become an issue and focus for environmentalists, and for environmental and health administrators. Irrespective of scientific evidence to support a benign impact of wood preservatives on the environment, both public and political pressure will probably demand the use of less-toxic chemicals. The dilemma of finding and introducing effective substitutes for the broad spectrum fungicides in current use is discussed. The role of new process development is highlighted in relation to providing "just-in-time" processing and integration of operations. The concept of totally contained treatment is demonstrated. In an era of product development, it is argued that the preservation industry needs to broaden its horizons to provide additional improvements to wood properties. These other properties could include dimensional stabilisation, fire resistance, and improved physical and visual characteristics.

2 INTRODUCTION

Technologies for reducing the incidence of decay in wood can be traced back to the ancient civilisations of the Egyptians and Chinese. However, from a European perspective, it was not until the 1830s that chemical preservative treatment of wood became established as we know it today. In 1838, John Bethell invented the "full-cell" process - a vacuum/pressure method of impregnating liquids into porous media such as wood. He also described a number of liquids suitable for improving the durability of wood - for example, water-based solutions of copper-sulphate, potassium chromate, borax, and oil-based preservatives derived from coal distillation. Creosote achieved rapid commercial acceptance, but it was not until the 1930s that the modern copper-chrome-arsenic (CCA) formulations were developed. Creosote and CCA together provide the backbone for wood treatment today, although two other forms of wood treatment should also be mentioned. These are boric acid/borax and Light Organic Solvent Preservative (LOSP) treatments.

Boron treatment became established in Australia and New Zealand in the late 1940s and the 1950s, primarily for the treatment of timber against insect attack. LOSP treatment systems became established in Europe (in particular the UK) during the 1960s following a slightly earlier introduction in the USA. LOSP used a water-insoluble, low-viscosity solvent as the carrier for fungicides and insecticides. Waxes and resins were also incorporated to improve water repellency and prevent preservative movement to the surface during solvent evaporation.

A large number of other preservative compounds have been introduced on to the market but have not gained acceptance either because of chemical toxicity, low efficacy, high cost, or corrosiveness. Examples of chemical compounds which have been rejected (on the basis of toxicity) include mercuric chloride, lead salts and cyanides.

The success of the long-established preservatives is due to their broad-spectrum activity against fungi and insects, apparent resistance to the build-up of preservative-tolerant decay organisms, and general robust character, resisting abuse from both treatment applicators and end-users. Low cost, ready availability, relatively low mammalian toxicity when "locked" into wood, and negligible environmental impact also contribute to their success.[1-2-3]

Many of the compounds of multisalt preservatives and oil- or solvent-based preservatives are intrinsically toxic and are coming under the scrutiny of environmentalists, and environmental and health administrators. The subject has become an emotive issue but one where polarisation of opinions should be avoided because the banning of these effective wood preservatives would have severe economic consequences.

3 THE NEED FOR WOOD PRESERVATION

At the first International Conference on Diffusible Preservatives held in Nashville in 1990, arguments were put forward for a dramatic rise in need for the preservative treatment of wood. The arguments presented by Vinden[4] included:

(1) The cost of decay and insect attack

The cost of attack from subterranean termites alone has been estimated to be $1.5 billion/annum in the USA.[5] Levi[6] has estimated that $2 billion are spent annually in the USA on repairing timber damaged by insect and fungal decay. The costs of remedial treatment in the UK are also very high. In NZ where there is mandatory treatment of building timbers there is virtually no remedial treatment industry.

(2) Trends towards the banning of soil poisoning

The problems associated with termite attack will be exacerbated with the banning of organo-chlorines as soil poisons and the substitution of less effective compounds or the banning of soil poisoning altogether. Maudlin[7] has stated that, in the absence of effective termite control measures, buildings in most parts of the USA will be attacked sometime during the life of the structure.

(3) Better use of scarce resources

Global concern over deforestation has never been greater and public opinion in many industrialised countries is demanding a halt to needless destruction of natural forests. However, availability of plantation timber is limited. Nicholas[8] estimates that properly applied wood preservatives will increase the service life of wood by a factor of at least 7. He estimates that, in comparison to the use of untreated wood, the added service life of preservative-treated wood provides an annual saving in the USA of $7.5 billion and an additional saving of $43 million annually associated with the growth of trees that would otherwise be harvested. Coggins[9] has made similar estimates in the UK where current levels of preservative treatment (24% of softwood imports - 2.1 million m^3) save an estimated 1.7 million trees each year.

(4) Building regulations and code reform

The world-wide adoption of performance specifications poses both a threat to and an opportunity for timber utilisation. The threat arises not from the philosophy of performance requirements but in establishing the effective service life of a component. Conservatism in minimising the life expectancy of a component does not reduce risk but increases the error in predicting costs in use and therefore the probability of specifying less appropriate materials. The opportunities arise from permitting the use of timber in multi-storey buildings (providing timber can meet demanding performance specifications).

(5) The low cost of preservative treatment

In New Zealand the cost of preservative treatment of all building timber as required by regulation is less than 0.75% of the cost of the house, and for this house-owners can expect freedom from insect and fungal attack for the life-time of the house.

(6) Value-added production at source

Dwindling wood supplies and therefore higher raw material costs will result in sawmills opting more for the production of value-added products and less for low-value commodity items. Increasing raw material costs will bring about a demand for integrated processing and "just-in-time" manufacture. Direct component sawmilling will become more widespread to improve sawmill productivity.

Product differentiation, particularly in the panel products industry, can be achieved very effectively through chemical modification. Durability is but one of the technological attributes required of a building component. Increasing attention will have to be placed on aesthetic characteristics, for example, yellowing and surface checking, longevity of surface coatings, and mechanical performance of the product. In many applications, chemical modification of wood could substitute for preservative treatment.

4 ENVIRONMENTAL QUESTIONS

Environmental pressure groups are questioning the toxicity of current preservative compounds and their impact on health, safety, and the environment:

(i) At industrial plants

(ii) During end-user handling

(iii) In use

(iv) After disposal of redundant materials
- chemical waste products
- demolition of buildings.

Progress towards resolving public concern has to tackle questions which are often difficult to quantify. A pragmatic approach might be:

- Is there a wood preservation environmental problem?
  - extent of proven risk
  - known effects
  - probable risks

- What is the context of risk?
  - industrial
  - public
  - health
  - environment

- What remedies are available?
  - alternative chemicals
  - alternative treatment techniques
  - methods of disposal

- What are the costs?

- What is the most effective method of containing risk?
  - legislation
  - regulatory inspection
  - codes of practice
  - training
  - chemical supplier monitoring of users
  - industry quality control
  - product specification
  - work methods

Public opinion has had a major impact on tropical forest utilisation despite fairly cogent arguments that devaluation of these forests through non-utilisation may accelerate their demise through burning for agriculture. It can be argued that, irrespective of scientific evidence to support a benign impact of currently used preservatives on health and the environment, there will be increasing public pressure to use less-toxic chemicals.

Environmental health and safety requirements will demand that:

(i) The preservative should be non-toxic to humans and the environment or, as a minimum, rendered non-toxic when fixed in the wood.

(ii) Treatment should be done when wood is in final shape and form to minimise generation of treated wood waste.

(iii) Plant operations should exclude emission of toxicants. There should be no soil, airborne, or waterway contamination.

(iv) Waste products (e.g., sludge) should be minimal.

(v) Redundant preservative-treated wood products should be disposable with minimum environmental disruption.

5 INTRODUCTION OF NEW PRESERVATIVES

The very superficial history of wood preservation given in the introduction highlights the use and success of broad-spectrum, fungicidally active chemicals. Traditionally, these have been applied as commodity treatments to broad categories of timber. The dilemma faced by the preservation industry is the difficulty of identifying alternative treatments which are as effective and robust as these traditional preservatives.

World-wide there is relatively little experience of the problems associated with attempting to introduce new preservatives. An attempt to introduce the so-called alkyl ammonium compounds (AAC's) into New Zealand in the late 1970s is quite well documented.[10] The chemicals were subjected to intensive biological testing and field testing to international standards, which probably constituted one of the most detailed evaluations ever conducted on new preservatives. Results from these tests concluded that some of these compounds were very effective biocides.

Four years after their commercial introduction, incidence of poor performance was reported. Reference to extensive commodity and graveyard trials established 7 years previously still indicated good performance and therefore it seemed likely that premature decay was associated with commercial application of the chemical.

After 8 or 9 years of field test exposure, some samples started to indicate superficial signs of decay. Approvals for using AAC's were rescinded by the regulatory authority.

The mechanisms of failure of commercial material are still not clear, despite extensive investigation. There are a number of theories:

(i) <u>Leaching characteristics</u>: Whilst the gross loss of chemical after leaching was the same for AAC's as for CCA, there were indications that the bulk of chemical loss of AAC's occurred on the surfaces rather than across the total cross-section. Such a difference could be attributable to differences in fixation mechanism. It is easy to postulate what impact such a leaching characteristic may have on preservative efficacy.

(ii) <u>Preservative distribution</u>: Fixation mechanism also has a pronounced effect on preservative distribution. The very fast fixation rates of AAC's mean that they are intolerant of poor treatment practices and plant. Steep preservative gradients in combination with machining after treatment will remove high concentrations of preservative and expose untreated wood. Exposure of such wood to high-risk environments (e.g., decking timber in close contact with soil) will inevitably lead to premature decay. AAC solutions can also "cream" with excessive agitation despite the addition of anti-foaming agents. Impregnation of wood with preservative/air mixtures will have a marked effect on the rate of preservative penetration.

Variation in pressure during flooding of the treatment vessel may also cause premature swelling of the wood surfaces. The permeability of wood is markedly affected by swelling and it is easy to postulate the potential consequences of poor plant control.

(iii) Build up of preservative-tolerant organisms: The surface active properties of AAC lead to rapid uptake of water by treated wood. The result is that AAC-treated wood remains at moisture contents suitable for decay development for much longer periods than CCA-treated wood which is partially water-repellent. This could well impact on the long-term performance of AAC-treated wood and is a factor that cannot be addressed from traditional testing procedures. The cause(s) of premature failure of AAC's are still a matter of speculation. However, the points illustrated above highlight a few of the potential problems associated with introducing new preservatives.

- None of the international standards used to assess preservative efficacy predicted the premature failure of AAC's.
- The efficacy of preservatives can be established with confidence only through long-term field testing.
- Preservative treatment equipment and practices utilised for one preservative chemical may not necessarily be suitable for the application of different formulations.
- New preservatives may be less robust or forgiving when abused either during treatment or in use if placed in a "high-risk" environment.
- The performance norm (expectation) has been established by a very effective preservative - copper-chrome-arsenic.

Research over a period of 30 years has failed to identify substitutes with such broad-spectrum applications as the CCA formulations. It seems inevitable that substitution efforts will need to be supported by more technical support to users and suppliers of treated timber and cope with a higher number of claims of "poor performance".

A disturbing trend in Europe is the acceptance of new preservatives without rigorous field testing. Clearly there are commercial problems in delaying approval of new preservatives until long-term graveyard or commodity tests confirm adequate performance. With AAC's, 9 years had elapsed before any problems were apparent in above-ground commodities. However, the fact that carefully planned, prepared, and documented field trials had been established and were available in advance of commercial use provided an invaluable resource from which predictions of behaviour in use could be made.

A protocol for approving new preservatives including the provision of field testing has been presented.[11] The provision of detailed field testing is argued on the basis of establishing a strategic resource for future prediction of performance — "an early warning system" — rather than as a test of efficacy prior to approval. Field testing can be established relatively cheaply and easily. Our policy at the Forest Research Institute is to establish such tests fairly early - soon after the preliminary screening tests.

CCA preservatives are employed in the total spectrum of decay hazards ranging from low hazard (insects) through to above-ground exposed to weather, in-ground contact, and marine. It is relatively easy to substitute CCA's in the lower hazard uses and the trend is towards the use of more specialised chemicals in niche uses. These chemicals will tend to be less robust than CCA and therefore greater expense will be incurred in providing more extensive technical back-up, to ensure proper application and use. It is envisaged that these costs will be off-set through marketing value-added environmentally friendly products and that price will be set through market valuation rather than the "cost-plus" of commodity treatment.

Given the public pressure for safer preservatives, one strategy is to use CCA more selectively for the more demanding uses where performance is more critical and where public access is restricted and environmental hazards minimal. Horticultural and agricultural uses come to mind. In extreme situations where leaching may be considered undesirable, the use of impermeable barriers on the outer surfaces of timber in ground contact must have merit.

There may be pressure to substitute inappropriate fungicides in some end-uses. I have heard of boron preservatives being recommended for exterior decking. This is an inappropriate use unless there is a suitable leaching barrier. In the end such recommendations will bring the preservative into disrepute.

Boron has been approved in New Zealand for 35 years for the protection of exterior sheathing, where a 3-coat paint system is maintained to prevent leaching. Such a system works and shortly polymeric barriers will be available for incorporation into the preservative to inhibit boron leaching.[12]

It is essential that standard test procedures imposed for evaluating the efficacy of a preservative system are realistically applied. It is known that standard leaching procedures will remove all boron salts from a cross-section, yet the preservative has achieved good performance when suitably coated. Leaching procedures using distilled water do not reflect the leaching hazards experienced by preservative-treated timber when placed in horticultural soils. These soils have high nutrient status, and have the capacity for ion exchange with preservative adsorbed on to the cellulose or lignin.[13] Thus, leaching tests should be carefully applied with regard to intended use, and should mimic the exposure hazard.

6 SOLVENT TREATMENTS

Light Organic Solvent Preservative (LOSP) treatment was developed in the UK during the 1950s and 1960s to provide treatment for machined timber without the undesirable side-effects of swelling and shrinking associated with waterborne preservatives. The industry grew rapidly and has been particularly successful for the treatment of high-value products like window frames where machine finish and dimensional tolerance are critical.

LOSP's were introduced into New Zealand in 1980. The high permeability of radiata pine and rigid process specifications frequently resulted in high solvent usage[14] and resin bleed of timber both at the plant and in use.[15] Excessive solvent on the surfaces of timber also led to poor paint primer adhesion and drying.[16] The treatment schedules used had a major impact on the extent of these problems. Rueping treatment of resinous heartwood samples could result in up to 63% of wood surfaces being covered with resin. Matched samples treated by Lowry and Bethell schedules resulted in 4% and no resin bleed respectively. The results of this work are illustrated in Table 1.

Table 1 Resin bleed after light organic solvent preservative treatment of resinous radiata pine heartwood

Process type		Timber drying method			
		High temperature		Conventional kiln	
		Preservative uptake	Resin bleed (%)	Preservative uptake	Resin bleed (%)
Bethell	x	70	0.0	24	0.0
	s.d.	16.6		6.5	
	CV%	23.7		27.1	
Lowry	x	55	3.8	27	0.2
	s.d.	12.5		7.6	
	CV%	22.7		28.1	
Reuping	x	98	62.8	35	12.5
	s.d.	18.9		5.4	
	CV%	19.3		15.4	

The table also highlights the impact of drying technique on heartwood permeability. High-temperature drying resulted in a 2-3-fold increase in preservative uptake compared to conventional kiln drying, whereas sapwood uptake remained essentially the same. The heartwood/sapwood preservative retention ratios in radiata pine are also influenced by process type. Typical ratios are 0.4 for Rueping and approximately 0.5 and 0.6 respectively for Lowry and Bethell schedules.

Research into paint primer application revealed that process type also had a major impact on the success of factory paint priming. Lowry and Rueping schedules often resulted in extended "kick-back" of solvent, resulting in poor primer application. Bethell treatment, on the other hand, facilitated immediate acrylic primer application.

In 1990 the New Zealand Timber Preservation Council introduced "results" type specifications for LOSP-treated timber of less than 25 mm thickness. This facilitated research into novel LOSP treatment systems and the introduction of a generically new process - "Sequential Vacuum Treatment".[17] Commercial trials with this process using radiata pine have demonstrated reductions in within-charge retention variability, a 30% saving in solvent usage, a 20% saving in treatment time, and a new plant design which has lower capital cost. More importantly, the surfaces of timber are left totally free of solvent and resin bleed and would probably be amenable to immediate painting with acrylic primers.

The "results" type specifications established for LOSP treatment define penetration (complete sapwood) and retention requirements.

Future NZ specifications will probably define minimum heartwood penetration requirements to ensure adequate in-service performance of heartwood. Heartwood penetration of many wood species is difficult to define because of very diffuse penetration through ray tissue. The durability of heartwood is also very difficult to relate to preservative penetration and poses a problem for researchers and specifiers alike. Ideally, research needs to establish probabilities of decay against treatment levels and decay hazards.

The imposition of minimum depths of preservative penetration in heartwood will pose difficulties for many treatment plants, particularly those using air-seasoning techniques. However, as already mentioned, results-type specifications will assist the innovation process, and there are a range of techniques available to improve heartwood permeability.

The construction industry world-wide is reforming building codes and defining performance specifications. The challenge is for mycologists to translate performance requirements into results-type specifications (as a means of compliance).

Solvent emission is vigorously controlled in the USA. Concerns over the "greenhouse effect" will probably result in similar controls in other countries. Sequential Vacuum Treatment provides incremental improvement in reducing solvent emission. The options for LOSP treatment are probably either solvent recovery or vapour phase application of preservatives — for example, vapour boron treatment.[14,18-19]

7 ELIMINATION OF TREATED WOOD RESIDUES

World-wide the residues generated by the wood processing industries have been estimated to total some 210-345 million m^3/year of which between 70 and 80% are generated by the sawmilling industry.[20] Machining of timber treated with waterborne preservatives has traditionally taken place after preservative treatment and drying so that changes in dimensions of timber during drying are accommodated and surface deposits of preservative are removed. The practice results in the generation of approximately 15% treated wood waste. In the past this waste material would have been raw material put into composite product manufacture – for example, particleboard or flakeboard. However, the imposition of health regulations is tending to prohibit the use of such waste materials because of the dangers of producing toxic fumes or dust during sanding operations. These restrictions have even been applied to boron-treated wood waste. The implications of these restrictions are:

(i) Treated wood waste now poses a disposal problem and cost.

(ii) Treatment of board materials is best done after all processing and finishing has been completed.

The feasibility of machining before preservative treatment with waterborne preservatives and redrying has been examined at the FRI, Rotorua. Preliminary results have been very encouraging.

The practice has been incorporated into the "Lite treatment process" - a new process resulting in a very low moisture content after treatment.[21] Subsequent redrying is also very rapid and even permits high temperature drying without the customary checking and collapse. Machining before dip/boron diffusion treatment is also now possible through the use of thickened solutions.

8 INTEGRATED PROCESSING

At present preservative treatment tends to be regarded as a discrete processing step, as is wood drying. This impacts on wood flows through processing plants and usually involves inefficient wood handling and a series of down-times as materials await each discrete processing step. Flow diagrams for current processing steps are compared below with predicted future processing steps. These opportunities are provided by new technologies such as green fingerjointing, vapour-boron treatment, the Lite process, etc. The objective is to provide a production-line approach, with the waste being eliminated prior to processing. The speed of processing also creates opportunities for "just-in-time" processing.

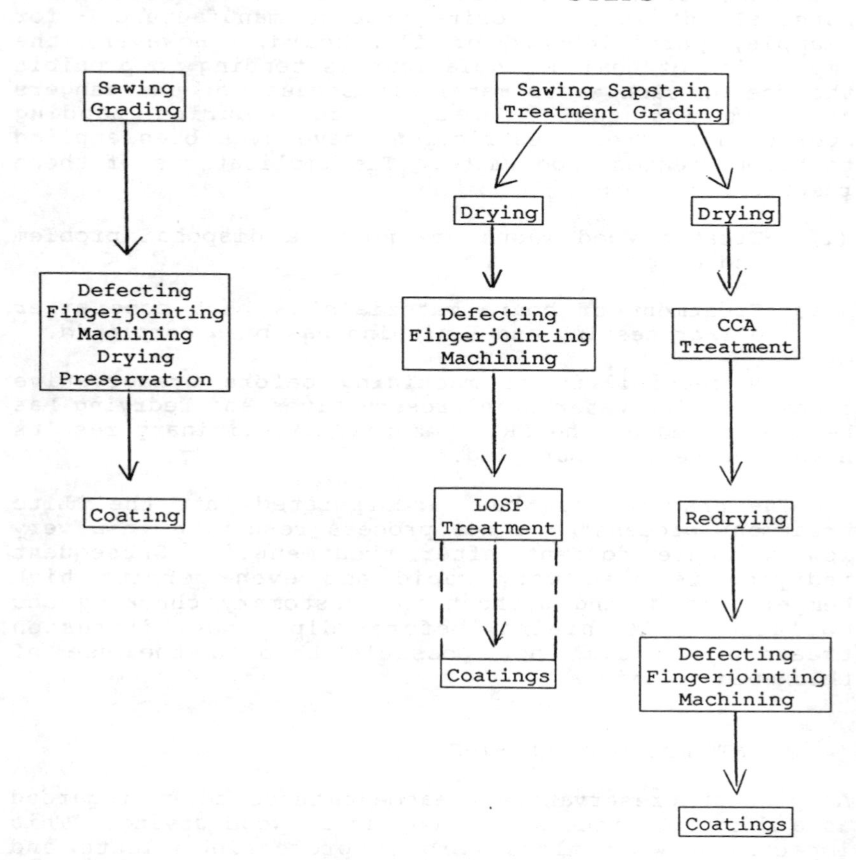

As long as preservative treatment is regarded as a discrete processing step, the impact of both down-stream and up-stream processing cannot be fully assessed. This can be overcome by an integrated processing approach, and it will also assist in developing the most cost-effective total processing system. Optimisation of individual processing steps may not necessarily lead to optimisation of the complete processing system.

9 ENVIRONMENTAL CONTAMINATION

The trend in new process developments is to provide environmentally friendly processes. Vapour boron, for example, is totally self-contained. There is no worker contact with the preservative. The timber is dry before treatment and dry on removal after treatment. Any residual vapour has been hydrolysed to solid boric acid and alcohol. There are no vapour emissions after removal from the cylinder because of solvent recovery. The concept of totally contained treatment will gain momentum in preservative treatment operations as a mechanism to achieve public acceptance.

Process automation and control is another feature of plant operation which is gaining momentum. Plant controllers have now been established for timber drying operations.[23] These provide "intelligent" control of banks of kilns — reducing operator error, providing modem linkages with both energy plant and remote operators, and warning of any potential problems. Automatic close-down of plant is also a feature of these controllers.

A system for intelligent control of preservative treatment plants is currently under development at the Forest Research Institute, Rotorua; the system will provide additional features such as an expert system for plant operators and a system of plant reconciliation which will provide automatic auditing of plant operations.

10 DISPOSAL OF REDUNDANT BUILDINGS

The disposal of preservative-treated timber once it has completed its useful life probably poses the most challenging problem for the preservation industry. Current practices in New Zealand recommend burial of treated timber in special land fill areas located away from waterways. The long-term fate of wood preservatives after burial has not been researched extensively and some concern has been expressed over the leachability of boron.

The impact of large-scale disposal of treated timber has not yet been experienced. A number of potential options are available:

(i) Recycling of timber
(ii) Reconstitution into other products
(iii) Chipping
- boiler feed for energy production followed by regeneration of any chemical residue and recycling
- slow-release micronutrient medium for forestry operations.

The first two options are not particularly practical. Work is in hand to investigate the potential for ashing and recycling of chemicals. The use of boron-treated wood chips as additives for boron-deficient soils is a practical option providing there is some rationalisation in the use of preservatives in building to avoid contamination with multisalts.

11 CONCLUSIONS

This paper has emphasised health and environmental issues because of the concerns being expressed by environmentalists and environmental and health administrators, and the impact these may have on our industry. The preservation industry needs to respond positively to these concerns and avoid polarisation of opinions.

The economic importance of wood preservation and its role in the environment is probably not well understood by the public at large. Attention has to be given to promoting the benefits of wood preservation. The historical significance and unique properties of CCA and creosote also need exposure.

Unfortunately, the wood preservation industry cannot be held up as an example of an industry concerned with either health or the environment, yet the underlying value of wood preservation is in resource conservation and improved material performance - to improve the quality (including health) of shelter, a basic requirement for human well-being. These qualities need to be part of the fabric of the preservation industry.

The industry needs to take stock of its performance and rectify its short-comings. Process development has reached a stage where industrial application of preservatives can be totally contained with no risk to worker health or the environment.

Public concern regarding the intrinsic toxicity of preservatives like CCA will continue. The loss of this preservative would be a major blow to the industry and has major economic ramifications. Even if scientific evidence supports a benign impact of these preservatives on health and environment, there may be value in restricting its application to specialty uses where demanding performance cannot be met by other less-toxic chemicals.

Substitute chemicals will tend to be less robust than CCA. It is therefore essential that long-term field testing is established prior to commercialisation. The temptation to promote substitute, non-toxic chemicals into end uses where they are clearly inappropriate should be avoided. However, the standard tests used to determine the suitability of preservative systems for specific end uses should realistically simulate the expected hazards in use.

Rapid changes are occurring in the wood processing industry which will have an impact on wood preservation or, more specifically, wood treatment. The changes are being driven by more difficult wood supplies (particularly tropical hardwoods) and higher price, less predictable markets, and higher stock-holding costs. These demand value-added production at source, integrated processing, just-in-time manufacture, automation, and control.

The wood preservation industry has an opportunity to broaden its horizons. Durability is simply one attribute of a product. Chemical impregnation can be used to impart other desirable properties such as surface hardness, dimensional stability, water-repellency, alternative material finishes, fire resistance, colour, and smell.

12 REFERENCES

1. R.D. Arsenault, Proc. Amer. Wood Pres. Ass., 1975, 71, 126.

2. R. Johanson and F.A. Dale, Holzforschung, 1973, 27(6), 187.

3. C. Grant and A.J. Dobbs, Environmental Pollution, 1977, 14, 213.

4. P. Vinden, Boron treatment in the 1990s, First International Conference on Wood Preservation with Diffusible Preservatives, Forest Products Research Society, Madison, Wisconsin, 1990.

5. H.B. Moore, Wood-inhabiting insects in houses: their identification, biology, prevention and control, USDA Forest Service and Dept Housing and Urban Development (Interagency Agreement IAA 25-75), Seattle, 1979.

6. M.P. Levi, A guide to the inspection of new houses under construction for conditions which favour attack by wood-inhibiting fungi and insects. US Dept Housing and Urban Development, OPDR, 1980.

7. J.K. Maudlin, 'Economic Importance and Control of Termites in the United States', USDA Forest Service, Southern Forest Experiment Station, New Orleans, 1986.

8. D.D. Nicholas and R. Cockcroft, Wood preservation in the USA, STU Info. No. 288, National Swedish Board for Tech. Development, 1982.

9. G. Shaw, 1990, Timber Trades Journal, 1990, 352(5900), 14.

10. J.A. Butcher, Proc. NZ Wood Pres. Ass., 1984, 24, 104.

11. M.E. Hedley and J.A. Butcher, Protocol for evaluating and approving new wood preservatives. International Research Group on Wood Preservation, 1986, Doc No. IRG/WP/2245.

12. P. Vasishth, R.C. Vasishth and D.D. Nicholas, The effect of boron addition on copper-wood equilibrium, First International Conference on Wood Protection with Diffusible Preservatives, Forest Products Research Society, Madison, Wisconsin, 1990.

13. P. Vinden, PhD Thesis, Imperial College of Science and Technology, London, 1983.

14. P. Vinden, Light organic solvent preservative treatment schedules for New Zealand-grown radiata pine, International Research Group on Wood Preservation, 1986, Doc. No. IRG/WP/3379.

15. P. Vinden, Resin bleed after light organic solvent preservative treatment: the effect of drying method and process type, International Research Group on Wood Preservation, 1986, Doc. No. IRG/WP/3378.

16. P. Vinden, Application of paint primers after light organic solvent preservative treatment, International Research Group on Wood Preservation, 1986, Doc. No. IRG/WP/3381.

17. K. Nasheri, P. Vinden, A.J. Bergervoet, and R.J. Burton, 'A Treatment Plant', New Zealand Patent Application No. 234588, 1990.

18. P. Turner, R.J. Murphy and D.J. Dickinson, Treatment of wood-based panel products with volatile borates, International Research Group on Wood Preservation, 1990, Doc. No. IRG/WP/3616.

19. P. Vinden, R.J. Burton and A.J. Bergervoet, Vapour phase treatment of wood with tri-methyl borate, Proceedings of Chemistry of Wood Preservation, Royal Chemical Society (in press).

20. G.K. Elliott, Uses and value of residues from sawn timber production, First ASEAN Forest Products Conference, Kuala Lumpur, Malaysia, 1989.

21. P. Vinden, A.J. Bergervoet, K. Nasheri, P. Cobham, N.P. Maynard and M.A. Brown, 'A Timber Treatment Plant', New Zealand Patent Application No. 234829, 1990.

22. P. Vinden, J.A. Drysdale and M. Spence, Thickened boron treatment, International Research Group on Wood Preservation, 1990, Doc. No. IRG/WP/3632.

23. W.R. Miller and S.G. Riley, Proceedings of the Wood Drying Symposium, Upgrading Wood Quality Through Drying Technology, F. Kayihan, J.A. Johnson and W.R. Smith (Eds), Seattle, 1989, p. 169.

The Properties and Performance of Coal–Tar Creosote as a Wood Preservative

W.D. Betts

TAR INDUSTRIES SERVICES, MILL LANE, WINGERWORTH, CHESTERFIELD, DERBYSHIRE S42 6NG, UK

INTRODUCTION

It is recorded that the earliest attempts to preserve wood from decay were in Egypt about 2000 BC where natural oils were smeared on the first wooden ploughs used for cultivation. In Roman times lees of oil and pitch were used for this purpose and since then many forms of protection were tried and used to prevent ships timbers from being ravaged by marine borers: cladding with lead, which proved too heavy and later with copper, which proved successful contrasted with the failure to remedy internal decay before the demise of wooden hulled ships.

The use of mercuric chloride in the eighteenth century by Homberg to protect timber against insects was the first practicable preservative. The Kyanizing process based on this compound proved to be effective in dry situations. However, together with copper sulphate and zinc chloride, the latter was used as a preservative for the first railway sleepers, these materials gave only short term protection in wet situations because of their solubility in water; the fixation of salts only became practicable a century later.

Coal was first carbonized on an industrial scale to produce tar as a substitute for wood tar used to preserve ships timbers and rope. However, the use of coal gas for lighting, attributed to Murdoch, resulted in the availability of by-product tar on a commercial scale and the setting up of tar distilleries to process it into distillate oils and pitch. An oil fraction, or more accurately a blend of oil fractions boiling substantially in the range 200 to 400°C is the creosote used for wood preservation. Julius Rütgers in Essen began to impregnate railway sleepers with creosote about 1846 using the process then recently developed by Bethell.

The value of wood preservation in this application became apparent with the passage of time: in present day terms, for the expenditure of 30% more for the cost of pressure creosoting based on the cost of the timber, the extension of its service life turned out to be impressive eg from three or four years for untreated beechwood sleepers to as much as 45 years for their preserved counterparts. It is a fact that in the UK when the navy no longer needed tar for wooden ships the infant tar distillation industry was saved from extinction by the urgent need to preserve the timber sleepers then being laid in enormous numbers from the middle of the 19th century onwards.

During this same period wood preservation by pressure impregnation with creosote was adopted for the treatment of railway sleepers throughout Europe, North America and subsequently elsewhere throughout the world; by the late 1950's it was estimated that about 2 billion creosote impregnated sleepers were in use throughout the world.

The extension of this method of preservation to other applications continued during the remainder of the 19th century: to marine and freshwater timbers, the poles required for the newly invented electric telegraph and later the telephone system, electricity transmission poles, wooden fencing and timbers for farm buildings.

Superficial treatment of timbers with creosote by brushing, dipping and spraying became popular for wooden garages, garden buildings, garden fences etc. because of the extension of service life achieved from 3 to 5 years, when untreated, to almost indefinitely by regular applications at a few years' interval.

Coal-Tar Distillation

As indicated above, coal-tar was a by-product from the carbonization of coal at high temperature, 1000 to 1200°C, for the manufacture of town's gas. Subsequently it was also recovered from the manufacture of metallurgical coke in by-product coke ovens and since the demise of coal gas manufacture in the 1970's is now the major source. In the 1920's the manufacture of smokeless domestic solid fuel began by the carbonization of coal at lower temperatures 600-700° from which a by-product tar is obtained but of lower aromaticity and higher phenols content than the previous two referred to above. In the UK only the process developed by Coalite has survived.

The crude coal tar is processed by continuous fractional distillation to produce an overhead light oil and usually four side-streams: naphthalene oil, wash oil, anthracene oil and base oil together with a pitch residue as the bottom product. The crude tar or coal oil as it is referred to from the Coalite process is distilled in a similar way to produce an overheads fraction, a middle oil, a heavy oil and a wax oil plus, again, a pitch residue. The analyses of two wood preservative creosotes are shown in Table 1.

Table 1 Typical Analytical Data Re-WEI Creosote Types A & B

Compound	% Mass WEI Type A	% Mass WEI Type B
Trimethylbenzenes	4.0	1.2
Indene and methylindenes	0.7	2.0
Naphthalene	8.7	14.9
Thionaphthene	0.4	0.6
Quinoline	0.5	0.7
Methylnaphthalenes	3.6	6.0
Biphenyl	0.8	1.1
Dimethylnaphthalenes	1.9	2.8
Acenaphthene	4.1	5.4
Dibenzofuran	3.3	3.6
Fluorene	4.1	4.6
Methylfluorenes	3.8	3.1
Dibenzothiophene	1.0	0.7
Phenanthrene	14.7	11.9
Methylphenanthrenes	1.0	0.7
Anthracene	1.1	1.5
Methylanthracenes	5.5	3.4
Carbazole	1.0	1.0
Fluoranthene	9.4	6.3
Pyrene	7.1	4.6
Benzo (a) pyrene	32 ppm	9 ppm
Phenols	$\leqslant 3$	$\leqslant 3$
Saturated hydrocarbons	<1	<1

Table 1 shows the presence of benzo(a)pyrene as a component but, as can be deduced from its boiling point of 496°C, is present only in small amounts: very much less than 0.1%.

Approximate physical properties of creosote are shown in Table 2.

Table 2 Physical Properties of Creosote

Property	Value
Boiling Point	approximately 200-400°C
Density	0.910-1.17 g/cm^3 at 25°C
Viscosity	4-14 mm^2/s at 40°C
Flash Point	above 66°C
Solubility in water	immiscible

Other properties:-

The effect of impregnation creosote (to BS 144) and creosoted timber on a range of metals, fibres, plastics and building materials has been studied. The study included observation of the weight change, tensile strength and crushing strength where appropriate. In brief, corrosive effects on a range of metals were slight e.g. for liquid creosote on mild steel a weight loss of 2.3 $\mu g/dm^2/day$; for creosoted timber a weight loss of 27 $\mu g/dm^2/day$ and both were significantly lower than the controls.

The effects on the breaking stress of fibres such as nylon and terylene were negligible and those for wool and cotton were increased.

The effect on concrete was only to retard slightly the development of the maximum strength of new material.

Whilst natural rubber, neoprene, pvc and polythene were significantly affected by creosote, Tufnol, PTFE, polypropylene and Perspex were least affected in that descending order.[1)] Fluorinated polyethylene is resistant to creosote and used as a packaging material for DIY sales.

An important practical advantage of creosoted timber is its low electrical conductivity; this is recognised in the use of creosote impregnated poles for electrical power transmission (an electrical resistance of 50-60 M Ω is typical for a pole) [2)] and, more importantly, for sleepers where track signalling is practised.

Some remarkable instances have been recorded of the fire resistance of creosoted timber. Mann [3] reported that a fire at the Maine State Pier at Portland, Maine, in October 1947 resulted in little damage to the pressure creosoted planking and piles of which it was constructed. He has recounted a similar case on the Pacific coast where the conclusion was reached that pressure creosoted piles are highly resistant to prolonged or intense heat. Similar testimony is given by Wilson[4] who described the case of a timber bridge fire in the USA and by Scott[5] who described the failure of deliberate attempts to ignite creosoted poles in South Africa. Van Groenou has summarized the above and other evidence rejecting the notion that creosoted timber constitutes a fire hazard.[6] The survival of pressure creosoted eucalyptus telegraph poles in contrast to similar untreated poles in a bush fire near Melbourne, Australia in 1962 was attributed to the thick smoke generated by the former from the creosote which shut off the supply of oxygen to the flames.[7] Finally, in a study of the ignitability of mining timbers, Demann and Herzig[8] found that the ignition temperature for creosoted timber was 50-100°C higher than that of untreated timber.

Fungicidal, insecticidal and biocidal properties of creosote

Amongst others Schulze and Becker[9] made an extensive study on a creosote, distillate fractions derived from it, the distillation residue and numerous individual components of creosote.

Efficacy against three fungi was studied using agar-wood block tests, in which creosote impregnated blocks are exposed to actively growing cultures. The results, expressed as toxic limits, are shown in Table 3. In the case of Poria (polyporus) vaporaria, it is obvious that the fractions boiling above 180°C have a greater efficacy than those boiling below. However, over the range 200 to 360°C the efficacies of the individual fractions are similar. Results for Lentinus lepideus (squamosus) show the same features. The results for Coniophora cerebella show fairly constant efficacy over the range 200 to 360°C.

The results of efficacy testing against insects on the same oil and fractions gave the results in Table 4 which show a high toxicity to the two species selected.

Table 3 Toxic limits in kg/m^3 of creosote oil fractions against fungi

Fractions °C	Coniophora cerebella	Poria (Polyporus) vaporaria	Lentinus lepideus (squamosus)
		Test fungi	
120-130 ...	-	>83.6	>83.7
130-140 ...	-	>72.0	>74.0
140-150 ...	-	>90.0	>92.0
150-160 ...	-	>88.0	>88.0
160-170 ...	-	>88.0	45.0-90.0
170-180 ...	-	>106	27.0-53.0
180-200 ...	5.40-7.80	8.4-13.5	8.6-12.6
200-220 ...	3.90-6.15	6.15-8.0	8.0-12.9
220-240 ...	8.20-12.6	6.45-8.8	12.9-16.4
240-260 ...	11.4-15.6	8.2-12.9	19.0-29.2
260-280 ...	6.15-8.40	8.8-13.2	20.5-32.2
280-300 ...	3.90-5.70	6.75-8.8	22.0-32.2
300-330 ...	3.90-6.15	12.3-17.2	20.5-30.8
330-360 ...	7.80-12.0	16.4-20.5	11.4-15.6
Residue ...	-	80.0-162	79.0-162
Original oil	5.10-6.80	5.55-6.60	24.6-30.0

NB The higher value indicates the smallest amount of preservative that still just protects the wood; the lower value indicates the largest quantity that is no longer able to give the wood adequate protection.

Table 4 Toxicity to insects kg/m^3

Fractions °C	Anobium punctatum		Hylotrupes bajulus	
	4 weeks	12 weeks	4 weeks	12 weeks
180-200 ...	>98	24-35	40-61	6-12
200-220 ...	36-57	23-36	3.2-5.5	<3.2
220-240 ...	21-36	9-15	6.0-9.0	3.5-6.0
240-260 ...	38-58	6-11	11-15	6.3-11
260-280 ...	27-42	8-12	12-18	12-18
280-300 ...	28-42	11-18	7.0-11	5.1-7.0
300-330 ...	>114	28-38	12-16	5.8-16
330-360 ...	>93	28-48	11-17	7.0-17
Residue ...	>98	>98	28-48	18-48
Original oil	60-110	33-54	14-32	14-32

Finally, the same authors carried out tests on the termite species Calotermes flavicollis. In this case wood blocks were impregnated with 10% solutions of the creosote or its fractions, then weathered for 1, 3 and 12 months before exposure to attack. All fractions boiling above 200°C imparted resistance to the blocks weathered for 1 month but those after 12 months only showed satisfactory resistance when the fractions boiling above 280°C had been used. Nevertheless, the UK Forest Products Research Laboratory (FPRL) recommend creosote as an excellent preservative against termite attack in their leaflet number 38 (1965) page 3 [10] providing the odour and staining associated with its use are acceptable. However, in areas of high termite hazard, e.g. regions of Australia, fortification of creosote with a termiticide is necessary (e.g. arsenic).

There is abundant evidence of the ability of creosoted timber to resist attack by marine borers such as Teredo and others. Many reports on the relative efficacy of various creosotes, their distillate fractions, residues and some of their individual components have been published. However, data on toxic limits is sparse. The degree of protection achieved depends on the level of impregnation. Very high retentions are necessary in timbers used in warm coastal waters, particularly where attack by Limnoria tripunctata is common, while lower retentions give resistance to Teredo. BS 5589 : 1989 for marine piling specifies retentions ranging from 180 kg/m^3 for larch to 400 kg/m^3 for beech.[11]

A suitable creosote should contain a high residue, boiling above 355°C; one suggestion is 40-50% as the optimum value. Too high a residue impairs its efficiency. A US Military Specification specifies creosote/coal tar where a solution of 70/30 is indicated.[12] A US Federal Specification requires 65-75% of creosote in the mixture.[13]

For situations of particularly high hazard, the use of additives to creosotes, such as copper and organolead compounds or even double treatments have proved successful.

Specifications for Creosote

Turning to specifications for creosote, most countries have their own: e.g. the AWPA P1 in the USA and BS 144 : 1990 in the UK in Table 5. In 1982 the West European Institute for Wood Preservation drafted a specification, see Table 6, which is gradually gaining acceptance throughout Europe. Two grades are specified: type-A mainly for heavy duty applications such as sleepers and marine timbers and type-B for all other uses. Variations are permitted in, e.g. crystallization temperature to accommodate regional differences in climatic conditions.

Scope and Methods of Application

Creosoting of timber.

Creosote can be applied by any of the known methods of timber preservation. However, in countries other than the UK application is gradually becoming restricted to methods of closed system pressure treatment. The objective is to achieve deep penetration of sufficient creosote to give the required long life to timber under the varied conditions of exposure.

Where these conditions are most severe, e.g. in fresh or sea water or in contact with the ground, pressure creosoting undoubtedly gives the best results. However, in many applications, where the environment is not so conducive to attack by the destroying agents, simpler non-pressure methods such as the hot and cold open tank process, dipping and brushing are also effective.

Before timber is used for construction work it should be seasoned, and this is of special importance where it is to receive any form of preservative treatment: to achieve satisfactory impregnation it is important prior to creosoting, to ensure that the timber is seasoned to a moisture content of below 30 per cent (the fibre saturation point) and there has been sufficient bacterial activity to increase the permeability of the timber to the preservative.

Obviously as much as possible of the cutting, framing and boring work on the timber be carried out before creosoting, for otherwise there is a risk of exposing untreated wood to attack by the wood destroying organisms.

Table 5
Proposed revision of British Standard Specification 144:1973 to be issued in 1990

Property	Type 1 for Sleepers etc		Type 2 for Poles primarily		Type 3 for brushing	
Liquidity. Temperature at which product is completely liquid after						
2 h, °C	32		32		–	
4 h, °C	–		–		0	
	Min.	Max.	Min.	Max.	Min.	Max.
Density @ 38°C kg/m^{-3}	1003	1108	1003	1108	–	–
Density @ 20°C kg/m^{-3}	–	–	–	–	910	1120
Distillation.						
Recovery @ 205°C % (m/m)	–	6	–	5	–	35
230°C	–	40	5	30	–	55
270°C	–	–	–	–	20	–
315°C	–	78	40	78	40	85
355°C	60	–	73	90	65	–
Extractable phenols content mL $100g^{-1}$						
Types 1 & 2 density range 1003-1045	5	20	5	20	–	–
1046-1108	0	20	0	20	–	–
Type 3	–	–	–	–	1	20
Examination of extractable phenols, mL 100 g^{-1} of phenol	–	–	–	–	–	4
Flash Point, closed, Pensky-Martens °C	66	–	66	–	66	–
Viscosity, kinematic @ 40°C, mm^2s^{-1}	–	–	4	14	–	–
Water Content, % (v/v)	–	1.5	–	1.5	–	1.5
Naphthalene Content, % (m/m)	–	–	8	25	–	–
Insoluble Matter Content, % (m/m)	–	0.4	–	0.4	–	0.4
Saturated Hydrocarbon Content, % (m/m)	–	15	–	15	–	25

Table 6 WEI Technical Specification

The two harmonized grades of impegnating oil are:-

WEI-Grade A : an impregnating oil consisting mainly of higher-boiling fractions;
WEI-Grade B : an impregnating oil consisting of medium-boiling fractions.

The impregnating oils must be pure coal-tar distillates without foreign constituents.

Characteristics	WEI-Grade A	WEI-Grade B
Density 20/4°C (g/ml)	1,04-1,15	1,02-1,15
Water content (vol.%)	max. 1	max. 1
Crystallization temp. (°C)	max. 23 *)	max. 23 *)
Content of acid constituents (vol.%)	max. 3 **)	max. 3 **)
Insoluble curve (vol%)	max. traces	max. traces
Boiling curve (vol.%)		
Distillate to 235°C	max. 10	max. 20
Distillate to 300°C	20-40	40-60
Distillate to 355°C	55-75	70-90
Benzo(a)pyrene content (***)	max. 1000 ppm	max. 50 ppm

* In southern European countries, such as Spain, Portugal, Italy and France, a higher crystallization temperature can be agreed between supplier and consumer, e.g. max.40°C for Grade A and max. 30°C for Grade B, as usual in these countries.

** Impregnating oils from low-temperature tar, produced in Great Britain, contain a higher content of acid constituents, but these have a composition different from those in oils from high-temperature tar. However, the water-soluble fraction is approximately comparable with that in oil from high-temperature tar. The above mentioned limit of 3% therefore does not apply to impregnating oils from low-temperature tar.

*** Some countries, (e.g. Austria, West Germany) classify products according to the concentration of benzo(a)pyrene.

Pressure creosoting methods include the Full Cell Process also known as the Bethell process, which leaves the cells filled with creosote and its main use is for timbers for marine pilings, exposed to the most severe conditions where heavy retentions are desirable. It is also necessary for timbers such as Spruce and Douglas Fir which are resistant to impregnation if an adequate measure of treatment is to be achieved.

Empty Cell Processes are the most commonly used pressure treatments, particularly where a clean finish is wanted. The Rueping and Lowry empty cell processes are used to obtain deep penetration of the timber without using excessive amounts of creosote. All creosoted transmission poles are by the Rueping process. The cycles for these three processes are illustrated in Fig. 1.

Open Tank Treatment

What is known as the 'hot and cold open tank process' gives excellent results in a relatively short time. The plant required is simple and cheap, consisting essentially of a suitably sized tank in which the creosote can be heated. The timber, weighted or clamped in position, is kept immersed in the creosote which is then heated to about 82-94°C, held at this temperature for a period so that much of the air in the wood is expelled and then allowed to cool and as a result of the partial vacuum created creosote is drawn into the wood.

The Butt Treatment Process is an adaptation for dealing with small numbers of fence posts, or similar pieces. The posts are stood upright in a good quality open ended drum containing creosote to a depth at least equal to that to which the posts are to be inserted in the ground. The creosote is then heated and allowed to cool as in the 'hot and cold open tank' process. The upper portions of the post can with advantage be given a good brush treatment at the same time with the hot creosote.

Brushing

One brush treatment cannot give the same length of life as the other methods but it can prolong the life almost indefinitely if repeated at intervals. Damp wood will not absorb the creosote so the work should be done

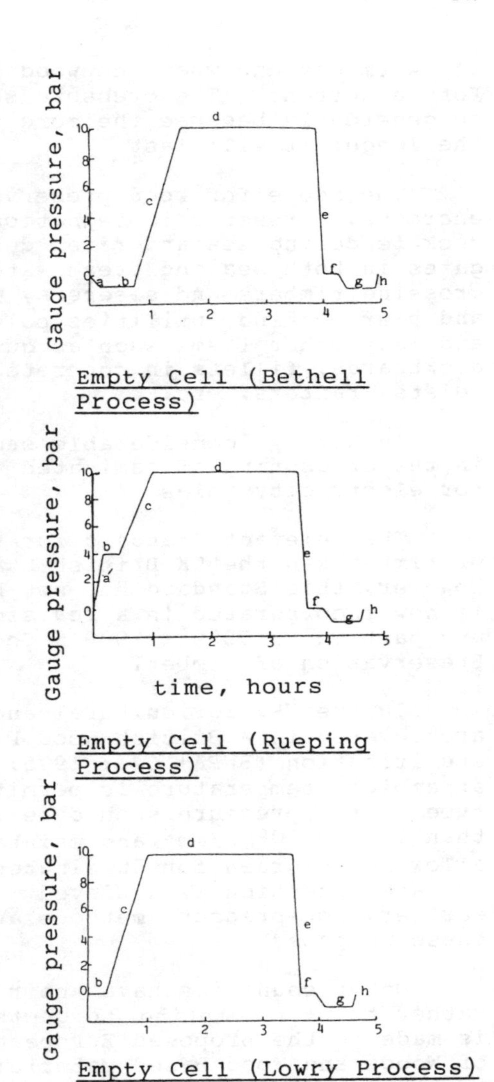

Empty Cell (Bethell Process)

Empty Cell (Rueping Process)

Empty Cell (Lowry Process)

a′) initial air pressure
a) preliminary vacuum
b) fill cylinder with creosote
c) build up pressure to maximum
d) pressing period
e) release pressure
f) empty cylinder of creosote
g) final vacuum
h) release vacuum

Temperature 65-100°C
Pressure 10-14 bar gauge

Comparison of treatment cycles for full-cell and empty cell processes

Fig. 1

on a warm day and when the wood itself is dry and therefore absorbent. The creosote should always be brushed on generously because the more that soaks into the wood the longer it will last.

The scope for wood preservation with creosote is enormous. Pressure impregnation is applied to piling, dock fendering sea and river defence works and lock gates in both sea and fresh water environments, railway crossing timbers and sleepers, motorway fencing, bridge and pier decking, utilities poles, cooling tower timbers and many other items such as guttering, troughing, duckboards, fillets in concrete, embedded timbers, joists, rafters, etc.

In Norway, considerable success has been achieved in the creosoting of laminated wood structures, e.g. for electricity poles.

The relevant Standard for the pressure creosoting of timber was the UK British Standard 913 : 1973. However, this Standard has now been superseded: part is now incorporated in a revision of BS 144 part 2 : 1990 and part in BS 5589 : 1989 a Code of Practice for Preservation of Timber.

In the UK, agricultural and horticultural timbers are covered in a British Wood Preserving Association specification (BWPA) C4 : 1975; pressure impregnation at ambient temperature is permitted using creosote to type 3; the pressure should be increased to not less than 1.03×10^6 N/mm^2 and maintained so that, after allowing to drain for 30 minutes, the minimum net average retention is achieved. And, as indicated earlier, non-pressure methods are still acceptable for these purposes.

Other countries have their own specifications. Rather than attempting to quote all these, reference is made to the proposed European Standard "Durability of Woods and Wood Based Materials" being prepared under the aegis of CEN/TC38 and when completed (presumably by 1992) will replace existing national standards.

It is general standard covering all wood preservatives. It is based on the concept of different hazard classes to which wood is exposed, five classes in all. It covers two broad groups of permeability, permeable and resistant and treatments required will be specified in terms of penetration levels (with appropriate analytical zones) and retentions. Thus, it is a 'results' type specification.

Some objections have been raised to the last because of difficulties in determining penetration achieved, mainly for the preservatives other than creosote.

Performance Records

Binns and Bennett in a paper to the BWPA[14] in 1951 described the excellent service record of Scots pine poles treated with creosote. They cited the case that some 8000 poles were still in service after 70 years. In 1957 Koppers Co. shipped three 63-year old telegraph poles from the UK to the USA for testing and the results were reported as good.[15] In 1982 Betts[16] reported to the BWPA the results of a study of a pole recently taken out of service erected in 1889 - the pole was well preserved and creosote remaining in the pole was demonstrated to be still an active fungicide. Webb et al reported on 83-year old poles in the USA.[17]

The excellent service of creosoted timbers in marine environments is recorded in the literature. I quote only two examples. Southern Yellow Pinewood piles treated with 380 kg/m^3 of creosote at Nueces Bay, Texas, survived for 30 years in marine borer infested waters and 90% was salvaged for further use.[18] After 28 years in the Long Wharf, Oakland, California, all the 1500 piles were reused. Those which could be observed were still in service after 34 to 45 years despite being exposed in marine borer infested waters.[19]

The AWPA regularly published full details of service records for railway sleepers and any randomly drawn example will confirm the claims for the performance of creosote as a preservative. A report on experimental sleepers after 44 years quotes that for creosote no life less than 40 years was recorded.[20] It should be noted that for sleepers in fast main lines mechanical degradation is believed to be a much more common cause of deterioration than fungal decay.[21]

The extensive field trials carried out over the period 1950 to 1980 under the aegis of the West European Institute for Wood Preservation are also worthy of note. These have been summarized and reported by van Groenou.[22]

Environmental Aspects

A paper by Anderson, Connell and Waldie[23)] presented to the British Wood Preserving Association (BWPA) annual convention in 1989, concluded that for wood preservation as a whole, overall the potential for a major risk is low, the number of instances of environmental damage are few and of small proportion; such damage as has occurred being generally restricted to the wood preserving site.

The areas of possible environmental problems are:-

1) air pollution by vapours or aerosols;

2) water pollution and soil pollution by leaching;

3) problems as a result of the exudation of creosote from impregnated timber, 'bleeding', causing environmental pollution and a hazard to both the industry's workers and the general public;

4) the disposal of redundant preserved timber.

Air Pollution

At ambient temperature the vapour pressure of creosote is very low, see Fig. 2. However, at the temperatures used in wood impregnation processes, 80-90°C, a small part, the lower boiling components can be liberated into the atmosphere. These components are substantially those boiling below about 280°, e.g. naphthalene, mono- and di-methyl naphthalenes and acenaphthene. According to their partial pressures the concentration in the vicinity of impregnated timber is very small. Moreover, their toxicological properties indicate that they are not dangerous chemicals in the working environment but do have a very low threshold of detection by smell, e.g. 0.0075 ppm in the case of naphthalene.

During the first 5 to 7 years after impregnation some 20 to 33⅓% of the creosote, the lower boiling components, are lost to the atmosphere by evaporation. However, the polynuclear aromatic hydrocarbons (PAH) which are the principal components lost to atmosphere are readily degraded by a combination of sunlight and air: the degradation reactions involved comprise

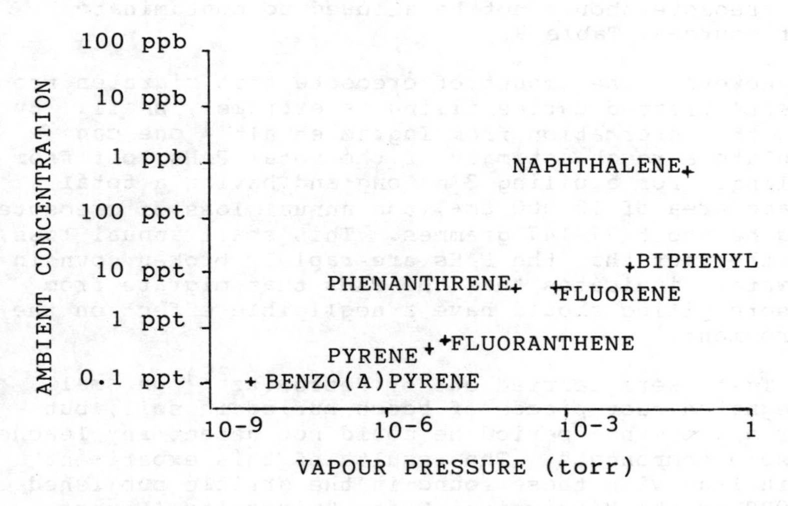

Plot of typical ambient concentrations of a series of aromatic hydrocarbons as a function of their room temperature vapour pressures.

Fig. 2

photooxidation, reaction with ozone and atmospheric pollutants such as oxides of sulphur and nitrogen. The half-lives of PAH under simulated atmospheric conditions have been studied: they are quite short and examples are given in Table 7.

Water and Soil Pollution by Leaching

Table 8 shows the solubility of some creosote components in water.[24] Although their solubility in water is low other considerations would indicate that pollution can arise from this source. Some ecological data concerning chemical and biological oxygen demands and various toxicity tests confirm this potential problem. Data presented obtained on the water extract from a light, brush application grade of creosote (referred to as carbolineum in Germany), demonstrates that creosote should not be allowed to contaminate water courses, Table 9.

However, the amount of creosote that migrates from creosote-treated marine piling is extremely small. By using the information from Ingram et al[24] one can calculate a rough estimate of the total PAHs lost from a piling. For a piling 3 m long and having a total surface area of 15 000 cm^2, the annual loss of creosote would be about 77-147 grammes. This small annual loss, plus the fact that the PAHs are rapidly broken down in sea water, indicates that the PAHs that migrate from creosote piling should have a negligible effect on the environment.

Tests were carried out by Petrowitz[25] on small impregnated test pieces of beech buried in soil, but after a 3 months' period he could not detect any leached creosote components. The results of this experiment are in line with those found in the article published in 1988 by the Mississippi State University (Forest Prod. Utilization Lab.) in the USA [26] and results obtained in Sweden which showed that 10-20 mm from a creosoted wooden pole no contamination of the surrounding soil with components of creosoted oil could be detected.

Table 7

Half-lives of PAH under simulated atmospheric conditions (expressed in hours)

	Simulated Sunlight	Simulated Sunlight + Ozone (2 ppm)	Dark Reaction Ozone (0.2 ppm)
Anthracene ...	0.20	0.15	1.23
Pyrene ...	4.28	2.75	15.72

Table 8

Solubility of PAH from Whole Creosote in Fresh Water and Sea Water*

Compound	Fresh Water (ppb)	Sea Water (ppb)
Naphthalene ...	3342 ± 452	1158 ± 371
2-Methylnaphthalene ...	1151 ± 109	784 ± 163
1-Methylnaphthalene ...	982 ± 80	784 ± 123
Biphenyl ...	196 ± 7	155 ± 27
Acenaphthylene ...	256 ± 69	238 ± 3
Acenaphthene ...	688 ± 88	640 ± 107
Dibenzofuran ...	482 ± 40	463 ± 47
Fluorene ...	405 ± 33	388 ± 44
Phenanthrene ...	660 ± 30	620 ± 70
Anthracene ...	168 ± 23	135 ± 21
Carbazole ...	368 ± 19	207 ± 35
Fluoranthene ...	184 ± 36	124 ± 20
Pyrene ...	103 ± 28	59 ± 6
1,2-Benzanthracene ...	43 ± 20	25 ± 12
Chrysene ...	36 ± 12	9.7

*Approximately 150 ml of creosote oil was stirred in 2 litres of fresh water or sea water for three days at temperature of 22°C

Table 9

Important ecological data of carbolineum (determined from an aqueous extract)*

Kind of determination	Values
Decomposition behaviour	
COD (Chemical oxygen demand)	300-330 mg/l
BOD_5 (Biological oxygen demand)	230 mg/L
TTC-test (activity of dehydrogenase)	$\leq$ 1:10
Fish toxicity	
LC_{50} (leuciscus idus aelanotus, 48 h)	50-100 mg/L
Daphnia toxicity	$\geq$ 1:64

*Willeitner, H. and Dieter, H. O. (1984) Steinkohlenteerol Holz als Roh-und Werkstoff, 42, 223-32

Problems as a result of exudation of creosote from impregnated timber (bleeding)

The AWPA Manual of Recommended Practice contains a glossary of terms in which 'bleeding' is defined as 'the exudation and accumulation of liquid preservative on the surface of the treated wood'.

Generally most workers have come to the conclusion that a combination of factors can give rise to an increased propensity to 'bleed'. Two of these factors are considered in what follows:-

The character of timber in relation to the resistance it offers to impregnation by creosote appears to be an important factor. Firstly, different species of timber show a different permeability to creosote due to differences in their structure, e.g. spruce which is classed as resistant versus pine which is classed as permeable. Secondly, the structure in the timber can be modified by prolonged contact with water thus making the timber, after subsequent drying, more permeable to pressure treatment with creosote; Holmgren noted, amongst others, that timber transported by flotation is much more permeable to creosote and less likely to 'bleed' than timber which has been transported by road.

The answer to the 'bleeding' problem in poles thus appears to lie in flotation or ponding the timber to be treated. Unfortunately, flotation of timber is on the decrease, due to increasing transport by road haulage, and ponding or accelerated treatment with bacteriologically active water by e.g. the FPRL method do not appear to be economically attractive to the creosoter.

A contributary factor may be the viscosity of the creosote oils used in some cases. It would be expected that the flow of creosote through the capillaries in timber being treated obeys Poiseuille's law, and on this basis the dependence on viscosity (n) of oil is apparent: the lower the viscosity the faster the flow. The viscosity of creosote is, of course, temperature dependent: the higher the temperature the lower its viscosity. Thus to ensure that emerging from the pores in the wood does so quickly and leaves a clean surface it should be as hot as possible. Although the oil in the impregnation cylinder is hot - between 80 and 90°C the timber charge does take a considerable time to heat up to the oil temperature and thus creosote inside the timber can be considerably cooler and hence more viscous.

Thus, some improvements in creosote impregnation and a reduction in bleeding propensity have been achieved by arranging to pre-heat the timber in the impregnation cylinder with creosote prior to applying the impregnation pressure. The well-known Boulton process (developed in 1879) which is still used in the USA, involves keeping the timber in hot creosote under vacuum for several hours. By heating under vacuum part of the residual moisture present is also removed and this makes the timber more suitable for treatment. The Deutsche Bundesbahn recommend circulation of hot creosote through the timber in the cylinder prior to impregnation, especially for Beech sleepers. This procedure improves the heat transfer to the timber resulting in improved penetration and recovery of more creosote oil (20 to 30 kg/m^3) under the final vacuum applied in the Rueping process.

A recent paper by D. de Jong[27] on new aspects of wood creosoting with respect to process and environmental aspects considers all the foregoing features with regard to pollution. The author advocates an after treatment of the freshly impregnated wood before removal from the plant as the main remedy: thus the impregnated material is transferred to an after-treatment vessel and treated with steam to remove all surface creosote. The condensed steam and creosote mixture is then transferred to storage and separated subsequently.

A paper by Kuijvenhoven[28)] presented in con-unction with the above by de Jong, describes a recently developed process "The Mark-Ecosafe R System MES" which he described as unique and a new way of impregnation with creosote oil to give maximum durability of the product combined with minimal environmental pollution. This new system is in two parts:-

1) modifications to the creosote treatment vessel and facilities;

2) a system for after treatment of the creosoted timber.

Using this system, wood can be treated in the traditional way but with the following modifications:-

1) programmable logic computer (PLC) control of the proces variables at the impregnation stage to avoid mal-operation;

2) the impregnation vessel is erected on "load cells" so that accurate retentions can be achieved and their achievement monitored during the process, and not afterwards as is the current practice. In this way desired retentions are not exceeded;

3) a hot storage tank replaces the Rueping tank and in this water and light oils can be vaporized off (then condensed and collected);

4) vapours in air from the cylinder and the hot storage tank are taken to a buffer vessel, to prevent large pressure variations in the system. They are then released and condensed and the waste gas burnt in the boiler house;

5) the treated timber is then charged to the after-treatment vessel and contacted with steam to remove surface creosote. The con-densed steam and creosote is then pumped to

a collection tank;

6) creosote and water from the collection tank are separated and the water passes to a waste water treatment system before discharge.

The author claims that compliance with quality assurance requirements can be achieved with the aid of the PLC system. This process, as outlined in the paper, appears capable of meeting all the environmental requirements.

Redundant Material

Creosoted timber can be used after its first service life, e.g. as railway sleepers, for another long period of several decades for a second period service, e.g. for fencing or in landscape construction.

Finally redundant creosoted timber can be burned in suitable installations. The calorific value of creosote (8 600 kcal/kg i.e. an additional 130 000 kcal per impregnated beechwood sleeper) will contribute to the energy recovered by incineration of the used wood. The products of the combustion of creosote are carbon dioxide and water vapour, thus completing the general natural carbon cycle.

In this connection Koppers Inc. have developed what they refer to as the Coregeneration process. For example, disused railway sleepers are combusted to provide heat for power generation - some of which is used in the treatment process to produce new creosote impregnated sleepers. The Deutsche Bundesbahn practises the burning of redundant railway sleepers in its power plant at Hanou.

Health and Safety Aspects

Coal-tar creosote has been used as an effective wood preservative for over 150 years without any significant health problems to either workers in the industry or the general public. Tests have been carried out on one commercially available creosote which indicated that it is practically non-toxic to rats following either percutaneous administration or oral consumption. It was also shown to be only mildly irritant to rabbit skin and practically non-irritant to rabbit eyes.

The International Agency for Research on Cancer, IARC, (1985)[29] concluded that there is limited evidence that coal-tar creosotes are carcinogenic to humans. Limited evidence indicates that a causal interpretation is credible but alternative explanations such as chance, bias or confounding could not be adequately excluded. Two extensive studies, both carried out by Henry in the UK, examined data from the period 1911 to 1938 on scrotal cancer and from 1920 to 1945 on cutaneous epithelioma. Since the early part of this century from which this data emerged, the practise of adding pitch to creosote has been discontinued and with the cessation of the manufacture of town gas from coal, creosote derived from that source of tar is no longer available. As a result, present day creosotes contain less of the higher boiling components and water soluble phenolic compounds. More recent data for cutaneous epithelioma from industrial sources show a very significantly lower occurence: in the UK 150 cases/a in the 1920-1945 period compared with 8 cases/a by the 1980s. In the earlier period the number of cases linked with exposure to creosote was 35 during the 25 years covered i.e. 1.4/a.

In contrast to the possible adverse effects, the beneficial effects of coal-tar products in human health care are well established as a dermatological treatment for psoriasis and related skin diseases. (Finser Inst. Copenhagen 1989, Stanford University School of Medicine 1977).

Finally, evidence for other forms of creosote-exposure related cancers in humans is sparse.

IARC also concluded the evidence for the carcino-gencitiy of creosote to animals is sufficient. However, many of the earlier studies were conducted and reported before modern scientific principles of carcinogenicity testing were developed and as a result were mainly of a qualitative nature and at toxic dose levels. For example from the results of Boutwell and Bosch (1958)[30] for creosote a dose for mice of about 40 g/kg is calculated.

Recent publications[31] cast considerable doubt on the use of results obtained from high-dose tests on rodents to determine the carcinogenicity of substances to humans (Ames and Gold 1990 and Cohen and Ellmein 1990) and the American Council on Science and Health has pronounced recently that the results of such high dose tests are being applied improperly to human beings who are exposed intermittently to traces of the substances in question.

To place these results in perspective reference is made to the classification of animal carcinogens by the American Conference of Governmental Industrial Hygienists (ACGIH); their classifications are for high, medium and low potency. However, those requiring doses greater than 1.5g/kg body weight to elicit a response in mouse skin are excluded classification because they are considered "not to be occupational carcinogens of any practical significance". The results by Boutwell and Bosch for the skin carcinogenicity of coal-tar creosote on this basis would be deemed as without practical significance from an occuptional exposure point of view.

The coal-tar industry believes that the risks from exposure to creosote are not significant in either its own or the wood-preservation industry. With regard to the general public and the amateur applier of creosote with a brush it is unlikely that they are exposed to anything approaching the same degree of risk as those employed in its industrial production and use. The creosote sold for amateur use is generally a lower boiling oil than that used for industrial purposes. Such an oil has recently been tested and shown not to be mutagenic by three procedures; the oil was also shown to contain less than 50ppm of benzo(a)pyrene, a substance which is becoming accepted as a marker of carcinogenic hazard. In the UK the Department of Health is content for the continued sale of creosote to the amateur under the Control of Pesticides Regulations of 1986.

These regulations which stem from the UK Food and Environmental Protection Act of 1985 require those involved in the bulk sale of wood preservatives to have the preservative approved by the Health and Safety Executive, to hold approval numbers and conform with regulated labelling requirements. Containers sold to the general public and contractors have displayed on them a list of precautions and advice for sensible use. If these are observed the health risks are negligible.

REFERENCES

1. W.D. Betts and F. R. Moore, paper to the British Wood Preserving Associaition Annual Convention 1978.
2. L. Borup, <u>Timber Technology</u>, March 1961, 69, 95-8.
3. R.H. Mann, <u>Proc. AWPA</u> 1950, 46, 339-46.
4. C.L. Wilson, <u>Rail Main. Eng</u>. 1925, 179.
5. M.H. Scott, <u>J. South African Forest Ass</u>., 1946, <u>13</u>, 21-34.

6. H. B., Van Groenou, Strassenbau u. Bautenschutz, March 1954, 3, 2-4.
7. Chem. Trade. J. 24.8.62, 151, 285.
8. Denmann and Herzig, Z.F. Bergbau, Eisen und Stahlind. March 1950, 15, Heft 3.
9. B. Schulze and G. Becker, Holzforschung, 1948, 2, 97-128.
10. Forest Products Research Laboratory (DSIR) leaflet No. 38, 3, 1965.
11. British Standard Code of Practise BS5589:1989
12. J.W. Roche, Proc. AWPA, 1961, US Military Spec. MLL-C-22599 (Rocks)
13. US Interim Federal Specification TT-C-00650 (AGR-FS).
14. R.G. Bennett and E.J. Binns, BWPA Annual Convention 1951, 37-41.
15. Private Communication, Koppers and Co Inc, Pitts, 19, Pa.
16. W.D. Betts, BWPA Annual Convention 1982, 10-25.
17. D.A. Webb et al, Proc. AWPA 1981.
18. A.P. Richards, Proc. AWPA 1952, 48, 268-70.
19. F.D. Mattos, AWPA 1919, 15, 2001.
20. Proc. AWPA 1956 Report of Committee U-3, Tie Service Record.
21. Dept. Sci. Ind. Res. Report of Forest Products Research (UK) 1953.
22. H.B. Van Groenou, Holz-Zentralblatt, Stuttgart, No. 48, 21.4.1980, 708-9.
23. D. G. Anderson, et al, BWPA Annual Convention, 1987.
24. L.L. Ingram et al, Proc. AWPA 1982, 78, 120-6.
25. H.-J., Petrowitz, Holz als Roh-und Werkstoff Springer Verlag 1989.
26. Mississippi State University Forest Prodcuts Utilization Laboratory, June 21, 1988.
27. D. De Jông, BWPA Annual Convention, 1988.
28. L.J. Kuijvenhoven, BWPA Annual Convention, 1988.
29. International Agency for Research on Cancer 'IARC Monographs on the Evaluation of Carcinogenic Risk of Chemicals to Humans, Vol 35, Polynuclear Aromatic Components Pt IV Bitumen, Coal Tar and Derived Products, Lyon, France, Jan 1985.
30. R.K. Boutwell and D.K. Bosch, Cancer Res. 18 1171-5 (1958).
31. B.N. Ames and L.S. Gold, Chem. Eng. News 1991, 7th Jan, 28-32.

Chemical Analysis in the Development and Use of Wood Preservatives

R.J. Orsler

TIMBER DIVISION, BUILDING RESEARCH ESTABLISHMENT, WATFORD, HERTFORDSHIRE WD2 7JR, UK

1 INTRODUCTION

Wood preservation is a much more complex process than might be envisaged, demanding much of those chemicals selected for this purpose. Such a selection is necessarily rigorous and depends on abilities to perform effectively and to endure.

The first stage in selecting a suitable pesticide is to ensure that it is capable of killing, or rendering harmless, the target wood-destroying organism. This target may be as selective as a single insect, but more usually is a group of insects or fungi. Biological testing systems can evaluate the effectiveness of the candidate pesticide and determine the quantities needed in the wood to ensure adequate protection. Although these tests do sometimes include in their procedures an acknowledgement of the method used to introduce the preservative into the wood, they mainly determine the quantities needed to provide protection, either assuming uniform loading throughout the wood substance (ie as toxic limits in kilograms per cubic metre) or when applied to the surface in measured quantities (grams per square metre). The tests are normally carried out on the sapwood of a standard timber species which is very amenable to treatment.

Timber is rarely penetrated uniformly by a preservative treatment. In gross terms the loading will depend on the proportions of sapwood and heartwood present, since sapwood is generally more treatable than heartwood, the severity of the method of treatment in forcing the preservative into the wood and, not least, on the timber species being treated. A range of methods exist which can lead to relatively shallow or deep penetration. However, even within the treated zone different quantities can occur in the different cell elements of the wood. Thus in attempting to assess the protection afforded by a commercial treatment on a particular timber species, information on the disposition of the preservative, obtained by chemical analysis, is most useful in relating the results of the treatment to the biological test data. Alternatively, if treated timber is exposed to service or field trials in order to assess its probable life, periodic chemical analysis, or analysis of failed samples at the end of the trial, provides important information on the persistence of the preservative system.

Treated timber is called upon to perform satisfactorily for as little as 5-6 years or as long as 60-100 years. During this period its surroundings can have a profound effect on its integrity. Changes in temperature, humidity and solar radiation associated with annual climatic changes can change radically the chemical composition of the wood and any associated wood preservatives. Treated wood exposed to rain or standing in fresh or salt water is subject to leaching, while that situated in warm, well-ventilated environments can lose relatively volatile preservatives by evaporation (sublimation). In such instances, a preservative treatment, initially seen as an effective system, can become changed to a point where it no longer provides adequate protection. Monitoring such changes through chemical analysis allows deficiencies to be recognised and rectified.

Where materials are used that are potentially harmful to persons or the environment, most present governments are concerned that they are properly controlled and economically applied. In the United Kingdom concern for control of wood preservatives is exercised through the Control of Pesticides Regulations (1986)[1] which inter alia contains the general requirement that a wood preservative must be approved as safe and effective before it can be used. Just as chemical analysis is used in conjunction with biological tests to determine effectiveness, so it is employed in association with toxicological tests to assess safety.

Chemical analysis is frequently mentioned in research papers on wood preservation, but all too often no source reference or description of the method used is given. This should not be encouraged, for frequently the soundness of the argument or hypothesis contained in the paper relies on the quality of the analytical results, and without examination of the methodology used the validity of the exposition cannot be assessed.

This paper gives a brief review of the methods of analysis used for the principal types of wood preservatives currently used in the United Kingdom. It includes also examples of the advantages of chemical analysis in providing a greater understanding of the effectiveness of wood preservatives, principally based on work carried out at the Building Research Establishment in the United Kingdom. Finally reference is made to the work that has been done in establishing British Standards in this area and emphasis is placed on the continuing need for methodology development, in particular as it relates to European initiatives on wood preservation standardization for the 1992 Single Market.

2 QUANTITATIVE ANALYTICAL METHODS

The principal types of wood preservatives currently used in the United Kingdom are presented in table 1. This demonstrates that the analyst in this field must deal with a wide range of chemicals including inorganic, organometallic and organic compounds. The variety of techniques used to determine quantitatively the active ingredients in a formulation or in treated timber reflects this necessary catholic approach.

Table 1 Principal wood preservative types used in the UK

Type	Examples
Tar oil	Creosote
Water-borne	Copper/chromium/arsenic Borates, Fluorides
Organic Solvent	Tributyltin oxide, Pentachlorophenol, Copper and Zinc Carboxylates, Gamma-hexachlorocyclohexane (Lindane), Synthetic pyrethroids

Tar Oil Preservatives

The tar oils, essentially creosote, were in use before the birth of modern chemical analysis. Their reputation and continuing use derive from a history of good performance over long service periods. They are extremely complex organic mixtures and therefore difficult to specify precisely. As exampled by the current British Standard, BS144[2], specifications usually rely on the definition of a few key characteristics with accompanying analytical methodology, eg naphthalene content, extractable phenol content, rather than exhaustive analysis. However, the development of capillary column gas chromatography, sometimes in conjunction with mass spectrometry, has allowed a much greater understanding of the composition of these materials[3].

Water-borne Preservatives

Water-borne preservatives are used for a range of end-uses. The most important of this group are the copper/chromium mixtures of which copper/chromium/arsenic is the most common. They have displaced creosote in many situations, partly from marketing activities and partly for economic reasons but also because they produce a cleaner treated surface and no lingering odour. They are, perhaps, unique amongst wood preservatives in that they are introduced into the wood as a water-soluble mixture of salts which then react in such a way as to render them water-insoluble. This enables the treated timber to be used in damp or wet conditions without subsequent loss of protection by leaching. Early analytical systems for timber treated with this type of preservative involved digestion techniques followed by colorimetric analysis. However, current methods employ a liquid extraction stage, using a sulphuric acid/hydrogen peroxide mixture, followed by quantitative analysis of the three key elements using atomic absorption spectrometry[4]. Alternatively, and especially for quality control work in industry, X-ray fluorescence analysis[5] is becoming increasingly common.

The other main water-borne preservatives, such as the boron compounds and the fluorides, remain water-soluble throughout their service lives and are therefore prone to leaching if placed in service environments where water can permeate the wood. However, to ensure their complete removal from the wood for analytical purposes, sodium hydroxide solution is used to extract them. Boron can then be determined by colorimetric analysis[6], as can fluoride[7], although for the latter the use of an ion-selective electrode is sometimes preferred[8].

Organic Solvent Preservatives

The organic solvent group of preservatives probably offers the biggest challenge to the analytical chemist, not least because a relatively diverse group of these materials is used in wood preservation. Table 1 lists as examples what might be regarded as the core materials of this group certainly for the preservation industry in the United

Kingdom. All of these compounds are introduced into the wood dissolved in a non-polar organic solvent and as such are deposited essentially in the free spaces within the wood (ie they do not penetrate the wood cell walls). They should therefore be easily removed by solvent extraction for analysis. As an exception to this, tributyltin oxide (TnBTO) can undergo chemical changes within treated timber during its service life and therefore requires an ethanol/HCl mixture for extraction to ensure total removal of the tin content[9]. The total tin content, often used to express retention of this preservative in treated wood, can be measured by atomic absorption spectrometry[9]. These chemical changes affect the performance of TnBTO for they involve the TnBTO molecule losing butyl groups to produce dibutyltin and monobutyltin compounds which are less fungitoxic[10]. Studies set up to follow this feature and to determine the nature of the organotin within the treated wood have been greatly helped by polarographic analysis which allows the simultaneous quantitative determination of these three very similar compounds from a single solution[11]. Other methods which have been used successfully to determine the various butyltins in treated timber include paper chromatography followed by atomic absorption spectrometry[12] and high performance thin-layer chromatography with scanning densitometry[13].

Lindane[14] and pentachlorophenol[15] are easily removed from treated wood by simple organic solvents, eg methanol, and can be determined quantitatively using packed column gas chromatography, although for some methods the phenol is easier to measure if first converted to the methoxy form by eg diazomethane. Copper and zinc carboxylates used to be relatively easy to analyse when only the naphthenates were used, the method relying on the removal of the metal ion from the wood with HCl followed by atomic absorption analysis[16]. However, as a consequence of recent developments other compounds have appeared on the market and one can now find carboxylic acid preparations, such as the acypetacs compounds, the versatates and the octoates, in addition to the naphthenates. This has introduced a new need to characterize the acid part of the preservative in order to identify the preservative in treated wood. Suitable methodology is currently being developed in the UK for a British Standard. A qualitative approach has been taken for the carboxylates, using fingerprinting identification techniques with capillary column GC, in association with the customary quantitative analysis of the metal ion present.

It is amongst the chemicals used as biocides in

organic solvent preservatives in particular that new compounds are appearing, in some instances due to environmental pressures. One such example is the increasing popularity of the synthetic pyrethroid insecticides, permethrin and cypermethrin now being widely used as replacements for lindane, especially in the remedial treatment field. These molecules contain asymmetric carbon atoms and thus present the opportunity for isomer mixtures. Current specifications for these compounds[17] include the requirement for particular isomer mixes, and therefore any analytical method must include the determination of the separate isomers. This has been successfully achieved using high-performance liquid chromatography where even separation into enantiomer pairs can be carried out using a chiral-phase column[18]. Gas chromatography has been widely used for the analysis of the synthetic pyrethroids. The earlier methods using packed columns were not without problems[19], but the later use of capillary systems seems to have resulted in greater accuracy and reproducibility[20,21]. This illustrates the need for detailed and thorough validation work in the development of analytical techniques for use in this type of area.

Compared with lindane and pentachlorophenol, the newest additions to the organic solvent group of preservative chemicals appear complex molecules; TCMTB, IPBC, azaconazole and dichlofluanid (figure 1) are examples. In some cases methods have been proposed by the manufacturers for analysis of their new formulations. However, the molecular complexity of these new compounds is such that the development of suitable analytical methods must be carried out cautiously and with an understanding of the chemistry involved. Much remains to be done in this area before analytical methodology is in place which will command wide acceptance.

3 EXTRACTION TECHNIQUES FOR QUANTITATIVE ANALYSIS

For most quantitative estimations the first stage is to remove the preservative from the wood by some form of liquid extraction. This procedure is vitally important since it must result in the total removal of the material to be analysed and it must be removed without unforeseen chemical changes. Extraction of tributyltin oxide (TnBTO) from wood treated with an organic solvent preparation of this compound provides an example. Table 2 shows the results obtained by the polarographic analysis of extracts obtained with different solvents, used sequentially on aged treated timber, and on TnBTO that had been exposed to UV

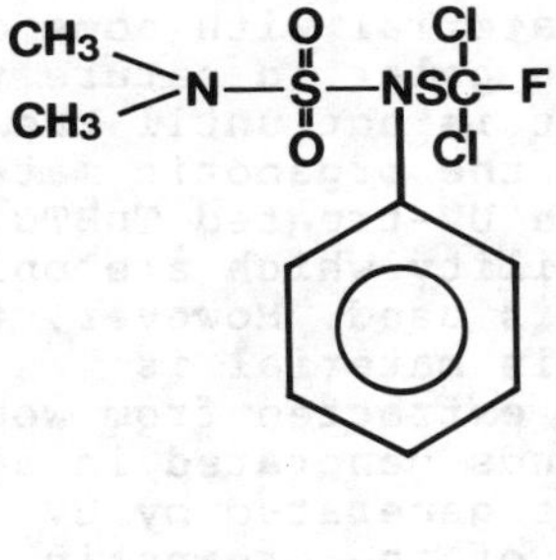

Dichlofluanid

IPBC

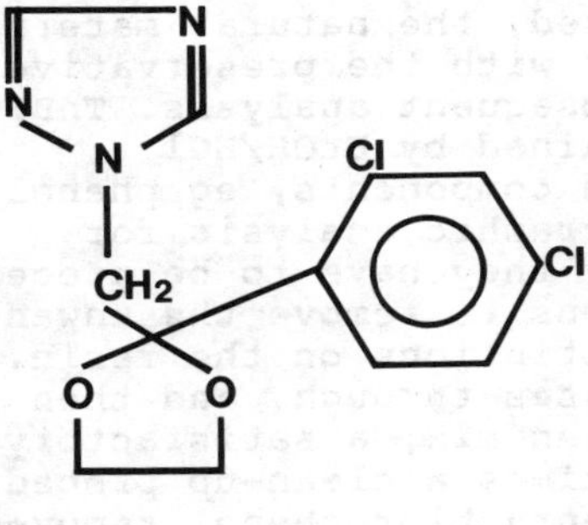

Azaconazole

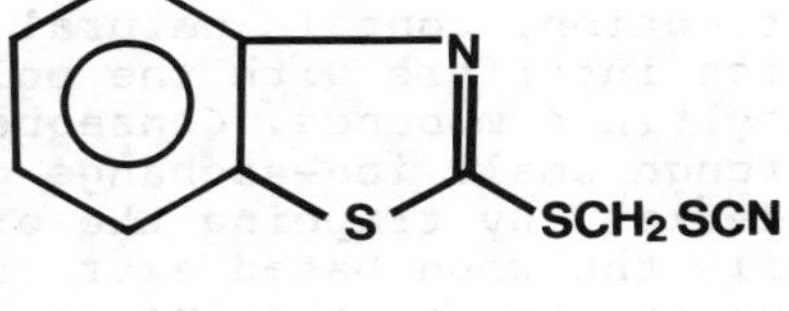

TCMTB

Figure 1 Examples of newer additions to the organic solvent group of preservatives

Table 2 Percent butyltins removed by sequential extraction from aged TnBTO-treated wood and from TnBTO exposed to UV light

Solvent	Treated wood			Exposed to UV light		
	Tri	Di	Mono	Tri	Di	Mono
Hexane	36	74	1	99	5	4
Ether	31	18	8	1	39	0
Acetone	18	5	39	0	23	25
Ethanol	15	3	38	0	33	71
Ethanol/HCl	0	0	14	0	0	0

light because it is well known that TnBTO degrades to lower butyltins when exposed to this type of radiation[22]. It can be seen from the treated wood example, that the use of simple non-polar organic solvents, such as hexane and ether, remove principally tributyl material with some di and mono. As the solvents become more polar in nature the di and mono content increases, but it is not until the ethanol/HCl mixture is used that all the organotin material is removed. Similar analyses with the UV-treated TnBTO also demonstrates difficulties with solubility which are only solved when the ethanol/HCl mixture is used. However, the pattern of tri:di:mono ratios for this material is different from that for the material extracted from wood, indicating that the organotin compounds generated in aged treated wood are different from those generated by UV light. Attempts to define the nature of the organotin compounds in treated wood have been carried out using Sn119 Mossbauer spectroscopy[23].

With some of the methods used, the natural materials of the wood are removed together with the preservative and these can interfere with the subsequent analysis. TnBTO extracts from treated wood, obtained by EtOH/HCl extraction, contain natural wood components, eg phenols, which interfere with the polarographic analysis for butyltin compounds. Consequently they have to be processed through small ion-exchange columns to remove the unwanted material. By trapping the organotin ions on the resin, while the wood-based extract passes through, and then removing the organotins as a clean mix, a satisfactory analysis can be completed. Sometimes a clean-up procedure is not necessary. Lindane and pentachlorophenol removed from treated wood by ethanol and subsequently analysed by gas chromatography require no clean-up because of the great dilutions involved and the use of an electron capture detector specifically incorporated for these chlorine-containing compounds.

As stated earlier, CCA preservatives undergo chemical change once in timber which renders them water-insoluble. This reaction is temperature and time dependent and it is clearly important to know when it is complete so that the treated timber can be released as soon as possible for safe and effective use. Prior to complete fixation the chemicals can be leached from the wood, allowing the opportunity for contamination of the surrounding environment and a reduction in the intended protection given to the wood. Water extraction of a series of matched, treated samples carried out at increasing intervals after treatment reveals the pattern of curing (figure 2). From this it can be seen that it takes approximately two weeks before the effect of water extraction is at a minimum. In addition, it shows that while the copper and chromium become virtually completely fixed, a proportion of the arsenic will

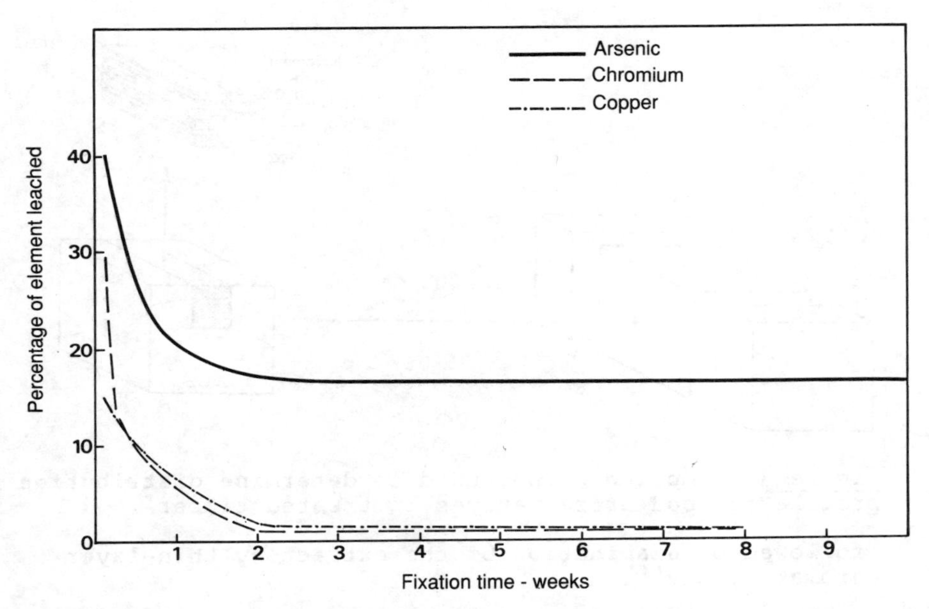

Figure 2 Percentage of elements leached from timber treated with a copper/chromium/arsenic formulation after various fixation periods at 20°C

inevitably be lost in service.

4 QUALITATIVE ANALYSIS

Qualitative analysis has an important part to play in the development of wood preservative usage. The level of penetration of a preservative into wood following treatment is extremely important, for the preservative must be distributed in such a way that it maximises its defence of the wood when wood-destroying organisms attack.The easiest way to reveal this is to spray the treated wood with a chromogenic reagent, having first cut the wood to expose an appropriate surface.For certain timber species spray reagents are also available which differentiate between sapwood and heartwood. Such simple methods can quickly provide much useful information, particularly where the disposition of the treated zone in relation to the heartwood/sapwood boundary is required. For the organic solvent preservatives, where it is sometimes difficult to discern which particular active ingredient has been used, characterization of the preservative can be achieved by liquid extraction of a small sample of treated wood,

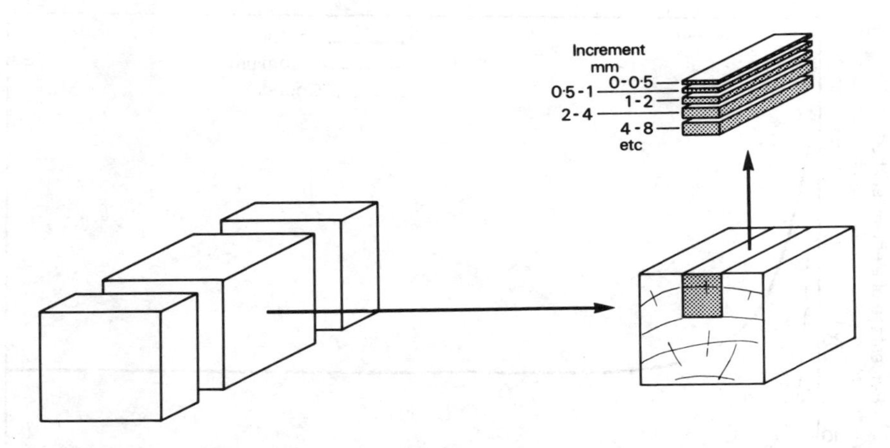

Figure 3 Sampling method used to determine distribution profile of wood preservatives in treated timber

followed by examination of the extract by thin-layer chromatography[24].

5 DISTRIBUTION STUDIES

As stated earlier, information on the distribution of the preservative in the treated wood is vitally important in assessing the effectiveness of the treatment. While chromogenic sprays show simple penetration patterns, they give no quantitative indication of the concentrations of active ingredients present. However, it is the concentration patterns within the wood (usually referred to as distribution profiles) which will indicate the effectiveness of the treatment. Furthermore, monitoring changes in the distribution profile with time has particular relevance to the study of the permanence of preservatives in treated wood. Several organic solvent preservatives have been studied in detail here because they can be lost from the wood by volatilization (ie they have small, but significant vapour pressures). To determine the distribution profile, a common procedure is to cut a sample block from the treated member, remove a representative piece of the face to be studied, and produce successive slices from the face into the block which can then be analysed as individual samples (see figure 3).

The results produced by using such a technique on a series of samples representing different exposure intervals, reveal how the distribution profile changes with time. In figure 4 it can be seen that lindane is lost from the outermost 2mm of the treated wood while deeper within the wood a more permanent situation exists. This type of information is extremely useful when used in conjunction with toxic values obtained in biological tests, for it allows an estimation of the amount of preservative needed in the wood initially, in order to ensure that sufficient remains in the long-term to maintain the required protection. This approach can equally be used in the development of accelerated ageing systems so that the long-term performance of new preservatives can be predicted from short-term tests. To do this the pattern of loss that occurs under normal conditions must be reproduced in the accelerated laboratory procedure. Monitoring the progress of this loss by chemical analysis allows a clear comparison to be made. Figure 5 shows the results obtained by conditioning treated wood in a vacuum oven. The similarity in pattern of loss with that shown in figure 4 indicates that this rapid ageing approach could be used in predictive testing.

The accepted standard method for evaporative ageing[25] utilises a the wind tunnel as a preconditioning procedure before biological testing. It has been assumed that such a procedure allows an estimation of the losses likely to occur through evaporation. However, in comparing the changes in distribution profile induced by the wind tunnel with those induced by the vacuum oven, important differences were observed in the analytical study (figure 6). The wind tunnel curves for TnBTO and tributyltin naphthenate both showed that during the conditioning period, which can be up to 26 weeks, a considerable amount of degradation had taken place. Thus this preconditioning procedure not only produced evaporative losses, but also reduced the effectiveness of the biocide by other means not intended in the conditioning procedure. On the other hand, the vacuum oven method produced little chemical breakdown but did promote the evaporative losses observed in practice.

6 STANDARDIZATION

For research purposes methods of analysis will be developed or selected to suit specific requirements. In such cases the methods need only apply to the subject under examination. However, where national or international standards specify levels of treatment for an assured performance of timber in service, generally accepted referee methods of analysis are necessary (a) to provide assurance to both producer and customer that the quality of

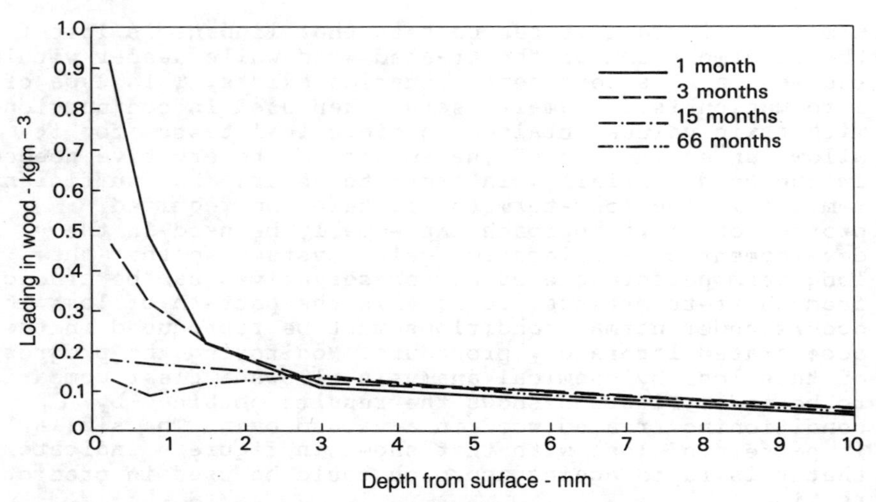

Figure 4 Changes in the distribution of lindane in treated wood with time under service conditions

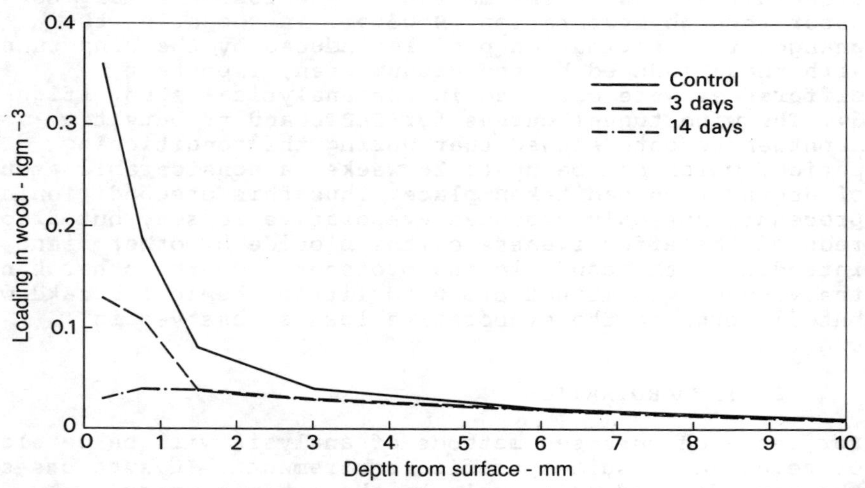

Figure 5 Changes in the distribution of lindane in treated wood with time when conditioned in a vacuum oven

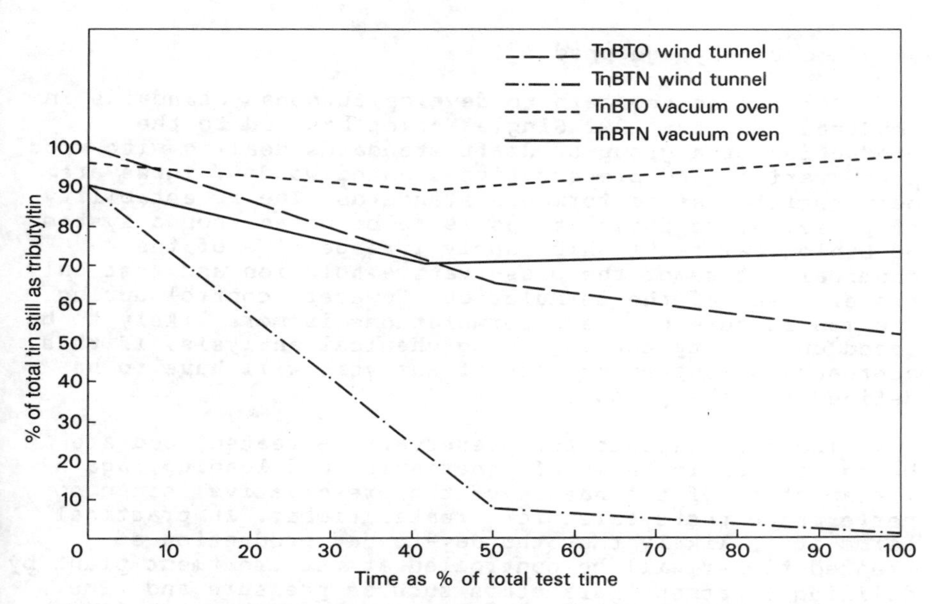

Figure 6 Comparison of tributyltin degradation arising from accelerated ageing in a wind tunnel and a vacuum oven

the product is as specified, and (b) to allow an independent analyst to determine treatment characteristics by a generally-agreed method in the case of dispute, or even litigation, over quality of treatment. The British Standards Institution have recognised the need for standard methods of analysis to be available by publishing the various parts of BS5666. These include methods for sampling prior to analysis[26], for qualitative analysis[27] and for quantitative analysis for copper/chromium/arsenic[28], copper naphthenate[29], zinc naphthenate[30], pentachlorophenol, pentachlorophenyl laurate, gamma-hexachlorocyclohexane and dieldrin[31], and tributyltin oxide as total tin[32]. Further parts describing methods for quantitative analysis for the various butyltin compounds, boron, permethrin and cypermethrin, and the other metal carboxylates are currently being prepared. British Standards are reviewed every five years when new methodology and changes in general practice are considered for inclusion. Each part of the BS5666 dealing with quantitative analysis contains two methods, one which is essentially an instrumental method and one which requires relatively simple laboratory apparatus. This approach is intended to permit any laboratory to carry out an agreed analytical method even if only moderately equipped.

The current concern to develop European Standards in readiness for the 1992 Single Market has led to the production of a group of draft standards dealing with wood preservatives and preservative-treated wood[33]. These are being written as performance standards. The acceptability of preservative formulations is to be based upon a system of biological tests which apply irrespective of the chemical nature of the preservative solution and test only the efficacy of the formulation. However, control during the manufacture of these formulations is more likely to be based on quality testing using chemical analysis. If this approach is adopted methods of analysis will have to be defined for the purpose.

The requirements for preservative-treated wood are being written in terms of penetration and loading, again irrespective of the nature of the preservative, since no performance tests exist for treated timber. In practical terms it is likely that the day-to-day production of treated timber will be controlled at the treatment plant by defining treatment parameters such as pressure and time. However, the initial claim that the specified treatment process produces the required penetration and loading characteristics will have to be demonstrated by generally agreed referee methods of chemical analysis. Again, periodic appraisal of the manufacturing (treatment) process can only be achieved by using those same analytical methods. These methods will have to be developed before such an approach can be implemented.

7 CONCLUSIONS

A range of analytical procedures, both qualitative and quantitative, has been established for determining the presence and distribution of wood preservatives in treated timber. In many instances, a critical part of these procedures is the method used to isolate the preservative from the wood. Chemical analysis provides a means of examining in some detail the methods by which the preservative chemicals may be introduced into the wood, and their subsequent fate during the service life of the treated component. Comparing the results from these analyses with the biological activity of the preservative formulations involved provides a clear indication of the performance that may be expected. The introduction of new preservatives and the increasing importance of efficacy assessment, particularly within the European Single Market, requires a continuing effort in the development of analytical procedures and the establishment of agreed methods within Europe.

REFERENCES

1. "The Control of Pesticides Regulations 1986", Statutory Instrument No 1510, H.M.S.O., 1986.
2. BS144 "Wood Preservation Using Coal Tar Creosotes, Part 1 Specification for Preservative", British Standards Institution, London, 1990.
3. W. Rotard and W. Mailahn, Anal. Chem., 1987, 59, 65.
4. A.I. Williams, Analyst, 1972, 97, 104.
5. A. Visapaa and P. Mansikkamaki, Paperi ja Puu, 1979, 4a, 253.
6. A.I. Williams, Analyst, 1970, 95, 498.
7. A.I. Williams, Analyst, 1969, 94, 300.
8. ion-selective
9. A.I. Williams, Analyst, 1973, 98, 233.
10. K. Nishimoto and G. Fuse, Tin and Its Uses, 1966, 70, 3.
11. Unpublished in-house methodology at Building Research Establishment, United Kingdom and ref. 9 above.
12. R. Hill, A.H. Chapman, A. Samuel, K. Manners and G. Morton, Int. Biodeterior., 1985, 21, 113.
13. S.V. Ohlsson and W.W. Hintze, J. High Res. Chrom. and Chrom. Comm., 1983, 6, 89.
14. R.J. Orsler and M.W.S. Stone, Int. J. Wood Pres., 1979, 1, 177.
15. R.J. Orsler and M.W.S. Stone, Mat. u. Org., 1982, 17, 161.
16. Unpublished in-house methodology at Building Research Establishment, United Kingdom.
17. Anon., "BWPDA Manual", British Wood Preserving and Damp-proofing Association, London, 1986.
18. G.R. Cayley and B.W. Simpson, J. Chromatog., 1986, 356, 123.
19. E.J. Kikta and J.P. Shierling, J. Chromatog., 1978, 150, 229.
20. J.F.C. Tyler, J. Assoc. Off. Anal. Chem., 1987, 70, 53.
21 P.D. Bland, J. Assoc. Off. Anal. Chem., 1985, 68, 592.
22. R.C. Poller, "The Chemistry of Organotin Compounds", Logos Press, London, 1970.
23. P.J. Smith, A.J. Crowe, D.W. Allan, J.S. Brooks and R. Formstone, Chem. and Ind., 1977, 874.
24. B.G. Henshaw, J.W.W. Morgan and N. Williams, J. Chromatog., 1975, 110, 37.
25. BS5761 "Wood Preservatives. Accelerated Ageing of Treated Wood prior to Biological Testing. Part 1. Evaporative Ageing Procedure", British Standards

Institution, London, 1989.
26. BS5666 "Methods of Analysis of Wood Preservatives and Treated Timber. Part 1. Guide to Sampling and Preparation of Wood Preservatives and Treated Timber for Analysis", British Standards Institution, London, 1987.
27. Ibid, "Part 2. Qualitative Analysis", 1980.
28. Ibid, "Part 3. Quantitative Analysis of Preservatives and Treated Timber containing Copper/Chromium/Arsenic Formulations", 1991.
29. Ibid, "Part 4. Quantitative Analysis of Preservatives and Treated Timber containing Copper Naphthenate", 1979.
30. Ibid, "Part 5. Determination of Zinc Naphthenate in Preservative Solutions and treated Timber", 1986.
31. Ibid, "Part 6. Quantitative Analysis of Preservative Solutions and Treated Timber containing Pentachlorophenol, Pentachlorophenyl Laurate, gamma-Hexachlorocyclohexane and Dieldrin", 1983.
32. Ibid, "Part 7. Quantitative Analysis of Preservative Solutions and Treated Timber containing Bis(tri-n-butyltin) Oxide: Determination of Total Tin", 1980.
33. Draft standards being developed by Technical Committee TC/38 of the European Committee for Standardization (CEN), Brussels.

Developing Quality Control Procedures and Standards for Diffusible Preservatives

Edward L. Docks

US BORAX RESEARCH CORPORATION, ANAHEIM, CALIFORNIA, USA

1 INTRODUCTION

This paper is based, in part, on a paper presented at the First International Conference on Wood Protection With Diffusible Preservatives in November 1990.[1]

In New Zealand, borate treatment has become the preferred treatment method for preserving timber in low-decay hazard situations which are defined as Hazard Class 1.[2,3] Since the mid 1950s, that country has been the leader in using borates as a wood preservative and currently they treat about 49 million cubic feet (1.4 million cubic meters) of timber with borates using a dip diffusion method. Other countries, such as; England, Malaysia, Canada, and Finland, have either used or are currently using borates for the same application. The efficacy of borates as a wood preservative has been extensively reviewed.[4] In addition, a review of the use of borates in the United States as a wood preservative has recently been published[5] as well as a review on the past and recent developments with particular reference to the use of borates as wood preservatives in the United Kingdom.[6]

Although borates have been used as wood preservatives in other parts of the world, use in the United States and Canada has been rather limited. Recently, there has been an increased interest in using borates for this application in the United States. The American Wood-Preservers' Association (AWPA), in the United States, is an organization consisting of wood treaters, wood researchers, chemical suppliers to the wood industry, and other individuals interested in the wood preserving industry. In order for wood preservatives to be included in the AWPA Standards, they must be approved by the appropriate preservative committee as well as the entire membership of the AWPA. In addition, a treating standard must also be approved by the appropriate treating committee as well as the entire AWPA membership. In 1989, sodium borates were approved by the American Wood-Preservers' Association (AWPA) as a wood preservative, and at the September 1990 AWPA Meeting a treating standard covering sodium borates was accepted by the appropriate treating committee. Full acceptance by the AWPA is expected in 1991.

The purpose of this paper is to discuss those elements which the author believes are important for the development of quality control procedures and standards for diffusible preservatives. Although sodium fluoride may also be considered a diffusible preservative, this paper will only concern borates.

2 DIFFUSION VERSUS PRESSURE TREATMENT

Currently, two principal methods are used to treat wood with borates. These are dip diffusion and pressure treatment. In the dip diffusion method green lumber having a minimum moisture content of about 40% (oven dry basis) is dipped for a few minutes into a concentrated borate solution, removed from the solution and then wrapped to allow the borate to diffuse naturally from an area of high concentration to one of lower concentration by means of the water contained in the wood. The diffusion process results in the distribution of chemical throughout the cross section after a period of a few weeks. As a general rule, 4 weeks of storage is needed for every inch (25 mm) of timber thickness.[3] If green lumber can be obtained, the diffusion process is especially suited for species which are difficult to pressure treat, such as Douglas fir and western hemlock.

In the pressure treatment method, lumber is placed in a pressure retort containing the treatment chemical, and treated using various vacuum/pressure schedules. For easily treated species, such as southern yellow pine, total penetration is readily achieved. For difficult to treat species, the vacuum/pressure times must be increased.

The treatment methods described above may require different quality control procedures as explained below.

3 QUALITY CONTROL IN THE WOOD PRESERVATION INDUSTRY

Analysis of the Treating Chemical

The quality of the treating chemical must be determined by chemical analysis even before treatment occurs. A procedure for determining boron trioxide in sodium borate wood preservatives has been published. This method is based on the mannitol titration procedure.[7] Determination of this oxide is an indication of the purity of the sodium borate.

Analysis of the Wood After Treatment

After the wood is treated samples must be removed and examined for penetration and retention of the treating chemical. The AWPA Standards contain a method for determining the penetration of borate preservatives.[8]

There are several methods which can be used to determine boron in wood. One method which can be used is to grind and ash the wood followed by extraction of the residue.[9] After extraction, the boron is in solution and can be analyzed. Since the amount of wood used for determining retention values is small (1-3 g), determining boron by titration is not the preferred method. However, recently a study has been reported where titration was used successfully to analyze boron in wood after extraction.[10] Trace methods, those which can be used to analyze <1000 ppm boron in solution, which have been used for boron include analysis using carminic acid, azomethine-H, atomic absorption (AA) spectroscopy, and inductively coupled plasma (ICP) spectroscopy. An overview of each of these methods is given in Table 1. The quantitation limit is the amount which can easily be quantitated. For the carminic acid determination,[11] the organics are eliminated during the ashing. Nitrate and nitrite have not been a problem in analyzing wood. The metal ion interferences for azomethine-H[12] are eliminated by using EDTA as a masking agent. High concentrations of sodium can be a problem when using ICP for analysis. This can be overcome by diluting the sample, if possible. The boron memory effect is the result of boron oxide forming on the ICP nebulizer when solutions of high boron concentration are analyzed. This is a particular problem when concentrated boron solutions have been analyzed, followed by analysis of a trace boron solution. In this case, a high boron value may be reported for the trace boron solution. The memory effect can be overcome by long rinse-out times between samples.[13] The limit of detection for boron in solution when using AA is about 50 ppm with a precision of plus or minus 25%.[13] Spiking solutions with known quantities of boron can be useful for determining interference

Table 1 Overview of Trace Boron Methods

	Carminic Acid	Azomethine-H	ICP	AA
Principle	Colorimetric	Colorimetric	Emission	Absorption
Quantitation Limit (ppm Boron)	0.5	0.5	0.1	50
Working Range (ppm Boron)	0-10	0-30	0-100	50-1000
Interferences	Organics, Nitrate, Nitrite	Cu, Fe, Al, Zn	> 1000 ppm Sodium, boron memory effect	None-Assuming a Stable Flame

problems which might occur with any of these procedures. Regardless of the procedure used to determine boron in solution, standard solutions should be included with every set of unknowns. Background boron values should be determined for untreated wood, if available, and then subtracted from the values obtained from treated wood. If an outside laboratory will be used to analyze boron in wood, that laboratory should be certified by submitting known boron standard solutions, as well as wood containing known amounts of boron. Analysis of these standards can indicate if the laboratory is capable of accurately analyzing boron in wood.

In order to determine the retention of borates in treated wood, initially, the borate has to be extracted from the wood. It is important to determine if most of the borate can be extracted. This was determined by pressure treating three 1′ x 1.5" x 3.5" (30.5 cm x 3.8 cm x 8.9 cm) southern yellow pine (SYP) boards with a sodium borate solution. The % boron by weight uptake can be determined by weighing the boards before and after treatment, multiplying this difference by the % boron in the treating solution, and dividing this result by the weight of the board before treatment added to the weight of boron resulting from the treatment. This result is referred to as the % boron by weight uptake. After treatment, the amount of boron was determined by ashing the wood and

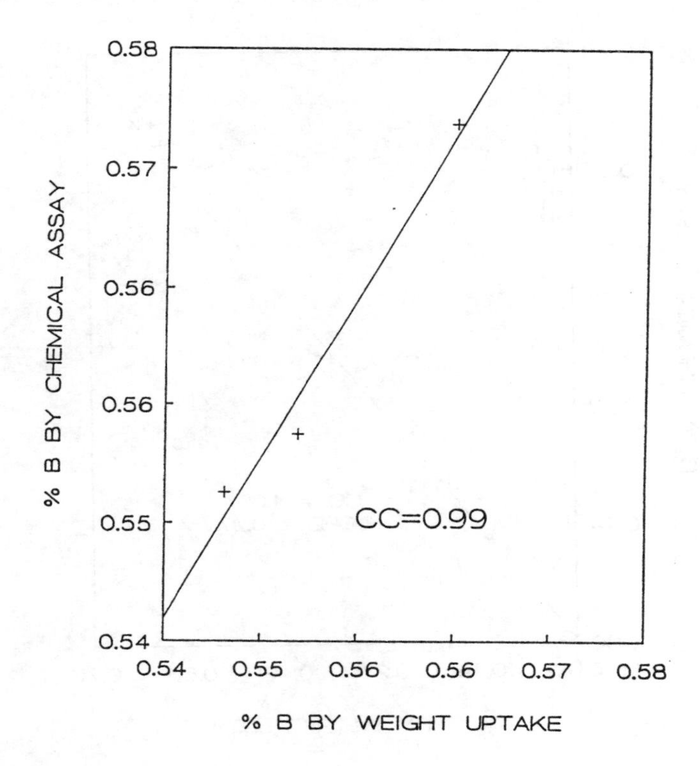

Figure 1 Percent boron in SYP 30.5 cm x 3.8 cm x 8.9 cm, weight uptake versus chemical assay

analyzing the resulting solution using the azomethine-H method as discussed above. These values are referred to as % boron by chemical assay. Figure 1 shows the results of this analysis graphically. It can be seen that the boron values determined by chemical assay correlate very well with the % boron by weight uptake. This indicates that most of the boron in the wood can be removed by the ashing procedure and quantitated using azomethine-H.

In an additional experiment, a number of 3/4" x 3/4" (1.9 cm x 1.9 cm) southern yellow pine cubes were pressure treated with a sodium borate solution. As described above, the % boron by weight uptake was determined as was the % boron by chemical assay. Figure 2 shows the results of this experiment. Again there is excellent correlation between the % boron by chemical assay compared to the % boron by weight uptake (correlation coefficient 1.00).

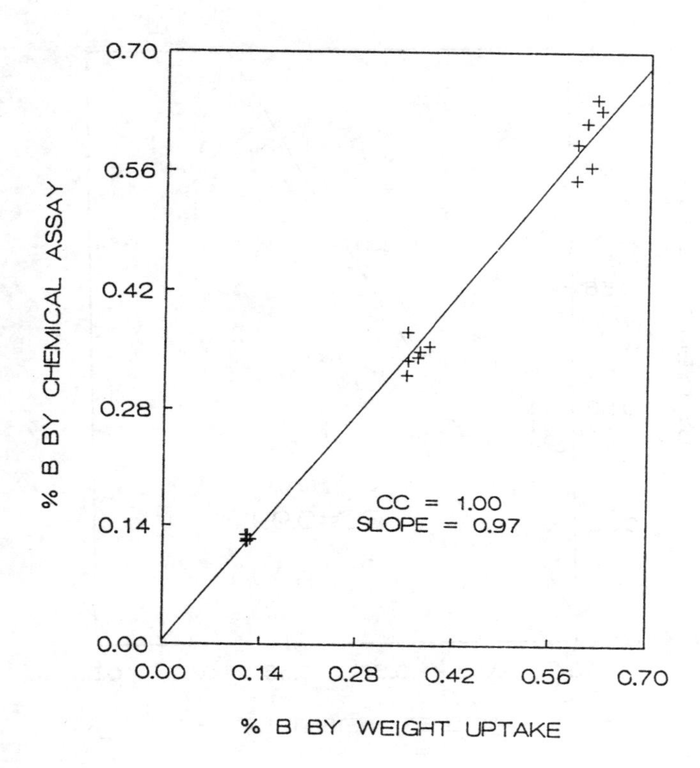

Figure 2 Percent boron in SYP 1.9 cm x 1.9 cm x 1.9 cm, weight uptake versus chemical assay

The slope of the correlation line also has to be considered. In this case the slope is 0.97 which indicates almost a 1 to 1 correlation of the values.

Although laboratories in countries, such as, England and New Zealand have been analyzing boron in wood for a number of years, this analysis is much less common in the United States. As a result of recent submissions to the AWPA related to a boron treating standard, it was decided to initiate a two part round robin test for determining boron in wood. In part one southern yellow pine blocks were pressure treated, to four retentions, by U.S. Borax Research Corporation (USBRC) using a sodium borate. The theoretical amount of boron was determined by the weight uptake technique described above. Cubes were sent to about 12 laboratories who were requested to analyze for boron by their preferred method. The results from each

Table 2 Summary of Part One of The Round Robin Evaluation

Lab	Method	Slope (Correlation Coefficient)	% of Chemical Assay Values Within 10% of Weight Uptake Values)
A	AA	1.33 (0.90)	17
B	Titration	0.97 (1.00)	92
C	ICP	1.04 (1.00)	83
D	Autotitrator	1.14 (1.00)	75
E	Azomethine-H	1.25 (1.00)	0
F	ICP	1.55 (1.00)	0
G	ICP	1.24 (0.99)	67
H	DCP	0.78 (0.96)	17
I	Azomethine-H	1.03 (1.00)	100
I	Carminic Acid	1.02 (1.00)	100

laboratory were returned to USBRC for comparison with the theoretical amount of boron in each cube. Results have been returned from 10 laboratories and these will be discussed below. The second part of the round robin test, which is just beginning, will involve the analysis of sawdust obtained by grinding a large number of southern yellow pine cubes which were pressure treated with a sodium borate to 6 different boron concentrations. The analysis will be performed by two methods which were submitted to the AWPA.[14] These are based on the carminic acid and azomethine-H methods. In addition, solution extracts obtained from borate treated wood will also be submitted to these round robin laboratories for determination of boron in solution. Results from the first part of the round robin test have been received and are summarized in Table 2.

Not only were a number of different methods used for analysis, but the method to remove boron from the wood was not the same for all laboratories. Some laboratories ashed, others extracted with various acids, acids and hydrogen peroxide, or hot water. The correlation coefficient for all of the laboratories was very close to 1, with the exception of laboratory A. This indicates that all of the values were very close to the regression line drawn when % boron by weight uptake was graphed against % boron by chemical assay. The slope for this line also has to be considered. The closer the slope is to 1.00 the better the correlation between % boron by weight uptake compared to % boron by chemical assay. A slope greater than 1.00 indicates that the % boron determined by chemical assay was smaller than the amount determined by weight uptake. This could be caused by loss of boron during the chemical assay or incomplete extraction of boron from the wood. Another criterion used to help determine how well these laboratories could analyze for boron in wood was the % of chemical assay values which were within 10% of the boron values determined by weight uptake. It can be seen from Table 2, that some laboratories (B, C, D, and I) did a competent job at determining the amount of boron in the cubes. Acceptable results appeared to depend less on the method used to analyze for boron and more on the laboratory doing the analysis. After results are obtained from the second part of the round robin test, those laboratories having difficulty analyzing for boron in wood will be contacted in order to help them with their analysis procedures.

4 QUALITY CONTROL IN THE TREATING PLANT

Quality control and inspection procedures in the treating plant have been specified by the AWPA[15] as well as the American Wood Preservers Bureau (AWPB).[16]

Inspection of Treated Timber Products

Inspection procedures for timber after treatment have been described by the AWPA.[17] A triple tier system is the preferred quality procedure. In this system, an independent inspection agency verifies the quality control performed by the treatment plant, and an enforcement body is responsible for monitoring the work done by the inspection agency.

5 CURRENT PENETRATION AND RETENTION REQUIREMENTS FOR FIXED PRESERVATIVES

In the United States the wood treating industry operates on results-type specifications. These specifications stipulate minimum retentions which are determined by chemical analysis in specified retention zones. They also stipulate process specifications which prescribe treating schedules to be used. For the most part, the specifications are contained in standards published by the AWPA.[18] In some countries a Hazard Class concept for wood preservation standards has been adopted. The Hazard Class system defines a number of different exposures for wood, each with a different degree of biodeterioration hazard.[2] However, in the United States, the standards as written by the AWPA, are currently based on the individual commodity, such as, poles, posts, plywood.[18] A portion of these standards are based on "fixed" type waterborne wood preservatives, such as chromated copper arsenate (CCA). As a result, the penetration for certain species is not expected to change once the preservative has entered the wood.

6 PENETRATION AND RETENTION REQUIREMENTS FOR DIFFUSIBLE PRESERVATIVES

For diffusible preservatives, such as borates, a standard different from that used for a fixed preservative may be necessary. For species which are easy to pressure treat, such as southern yellow pine, further diffusion after treatment will probably not occur. As a result, penetration of the borates will not change as a function of time. However, for species such as Douglas fir and western hemlock, where the borate may not fully penetrate after pressure treatment, additional penetration may occur as a result of diffusion. This was shown to be the case for freshly sawn boards of D. fir heartwood and western hemlock sapwood which were stored under a water spray prior to pressure treatment by full or empty cell methods.[19]

Therefore, for diffusible preservatives, the time to measure penetration and retention, for certain species, needs to be defined. Additional work is also necessary to determine the affect of moisture content on diffusion. Once this is established, standards for species difficult to treat by pressure processes, such as D. fir and western hemlock, can be written. If diffusion occurs, it may be necessary to define an assay zone which includes the portion of the wood containing the borate. This zone could be different depending on the species. A similar situation now exists in the AWPA standard for fixed preservatives.[18] For dip diffused lumber, the diffusion of borates will generally take longer than for pressure treated lumber. Therefore, an assay time would have to be chosen after diffusion has been completed. As a general rule, four weeks of diffusion, under cover, is needed for each 1" (25 mm) of thickness.[3] However this depends on several factors including the initial moisture content of the wood, species, sapwood/heartwood content, concentration of the treating solution, and diffusion storage conditions.

7 QUALITY CONTROL PROCEDURES AND STANDARDS FOR VARIOUS END USES

There are a number of end uses where wood treated with borates is already used or could be used. These include log homes, pallets, construction lumber and timbers. As each end use may require different species of lumber and/or treatment conditions, new standards may need to be developed for each of these uses. Standards for these industries are being developed in cooperation with the various end use manufacturers, builders, and architects.

Remedial treatments of wood already infested with pests using diffusible chemicals is another area of interest.[5] This use would require different standards than the one used for treatment of lumber.

8 QUALITY CONTROL AFTER TREATMENT

In the past there has been some reluctance to use diffusible preservatives, especially borates, because of a perceived "leaching problem". As a result, some users may feel that it is necessary to analyze borate treated wood which has been exposed to the weather prior to use. A number of papers, some as early as 1951, have been published which demonstrate that leaching of borates from treated lumber is relatively slow.[20-25] In one study[23] 1.5" x 3.5" (3.8 cm x 8.9 cm) D. fir lumber was pressure treated and exposed to simulated rainfall using a sprinkler. The wood was analyzed after various amounts of simulated rainfall. In accordance with AWPA standards for D. fir, the outer 0.6" of the wood was analyzed. The results of this test are shown in Figure 3. It can be seen that even after 88 inches of simulated rain, over 70% of the original sodium borate is present in the wood. As a result, it has been concluded that exposure of borate-treated wood to intermittent rain during transport or to occasional water contact once in use should not significantly decrease its efficacy.

9 OTHER BORATE WOOD PRESERVING STANDARDS

Up to this point, the AWPA wood preservation standards used in the United States have been described. However, there are other wood standards which include the use of diffusible borates. These have been published in New

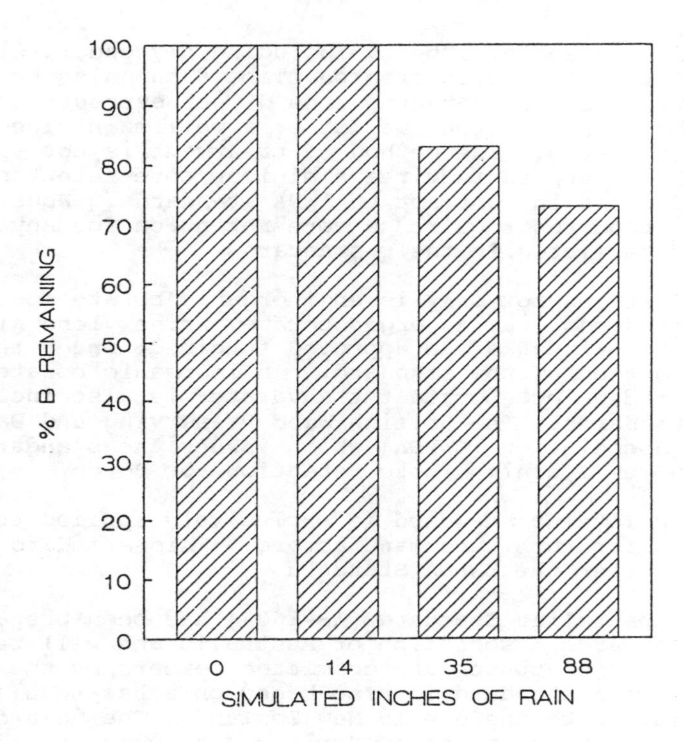

Figure 3 Persistence of boron in pressure treated D. fir exposed to simulated rainfall

Zealand,[26] Britain,[27] and Malaysia.[28] A draft standard has also been prepared by the Standards Association of Australia.[29] In addition, in 1992 the European Economic Committee (EEC) will publish a wood standard which will be used by its member nations.[30] This standard will be a results type specification with retention and penetration requirements.

In New Zealand, as mentioned previously, a hazard class system is used. In this system six categorized environments or conditions have been identified where timber is at a particular risk of biodegradation by one or more biological agents such as fungi, insects, bacteria, or marine organisms. Once the hazard is identified, the type of treatment for that particular hazard is given in the New Zealand Standard. The use of inorganic boron preservatives is only mentioned in the H1 hazard class.

This is for building timbers, including plywood, protected from the weather. Boron treated timbers can also be used for exterior use if protected from direct exposure of the weather by roof, external walls, or a well maintained 3 coat paint system. The method of treatment is not specified. This standard requires a minimum core retention of boron. This is in contrast to AWPA standards. None of the AWPA standards requires a core retention for any preservative listed in their standards.

In Britain, not only is an inorganic borate specified in the standards, but an organoborate, trihexylene glycol biborate, is listed as an approved fungicide under the biocide section of the standard. An inorganic borate rod, based on sodium octaborate tetrahydrate, is also included in the standards. The British Wood Preserving and Damp Proofing Association (BWPDA) which issues the standards, also requires a minimum core retention for Boron.

The Malaysian standard is exclusively applied to rubberwood for furniture manufacture. This standard is patterned after the BWPDA Standard.

The Australian Standard (AS 1604) has been prepared by the Standards Association of Australia and will be voted on by the appropriate committee members by the end of 1991. This standard is also based on a hazard class system similar to the one in New Zealand. The hazard classes where boron treated lumber can be used are similar but not exactly the same as those given in the New Zealand Standard.

10 OTHER QUALITY CONTROL CONSIDERATIONS

A number of other considerations should be addressed when developing quality control standards. Some of these considerations could pertain to fixed as well as diffusible preservatives.

1. New Zealand uses a hazard class concept for wood preserving standards where as in the United States, standards are written for an individual commodity. Should a hazard class concept be considered for the United States? This is currently being discussed within the AWPA.

2. The AWPA standards for retention are based on a weight/unit volume measure (lbs/cubic foot). However, it might be more appropriate to base retentions on a weight/unit weight basis. This concept has been presented to the AWPA[31] and will probably result in additional

discussions in the future.

3. A number of products containing borates are now being offered as remedial treatments for wood. These include borate rods, organoborates, and borate containing bandages for poles. Eventually quality control standards will need to be developed for these products. These standards should be published and reviewed. This could be done by the companies marketing these products and/or by an independent organization. This will have to be re-solved in the future.

4. Eventually, it may be necessary to have one set of standards which could be used on a worldwide basis. With the changes which will be occurring in Europe in the future, standards in the United States for preserved wood may have to more carefully match those written in other parts of the world.

These are a few of the issues which will need to be considered in the area of quality control for preserved wood.

11. CONCLUSIONS

This paper has addressed a number of items which should be considered when developing quality control procedures and standards for diffusible preservatives. Some of these issues will be the same as those required for fixed wood preservatives. For certain species of wood, the diffusion properties of these preservatives could determine the type of QC procedures and standards which will be needed to insure that the wood is adequately treated. Accurately analysising boron contained in the wood after treatment is very important. Various methods which can be used were reviewed. A brief summary of worldwide wood preservation standards, which include the use of diffusible borates, has also been discussed.

REFERENCES

1. E. L. Docks, Y. S. Chiu, and N. T. Carey, "First International Conference on Wood Protection with Diffusible Preservatives", Forest Products Research Society, Madison, WI, 1990, p. 115.

2. M. E. Hedley, Proc. - Annu. Meet. Am. Wood-Preserv. Assoc., 1990, 86, 31.

3. P. Vinden, J. Drysdale, and M. Spence, Int. Res. Group Wood Preserv., 1990, Doc. No. IRG/WP/3632.

4. R. Bunn, "Boron Compounds in Timber Preservation. An Annotated Bibliography -- Technical Paper No. 60", New Zealand Forest Serv., 1974.

5. H. M. Barnes, T. L. Amburgey, L. H. Williams, and J. J. Morrell, Int. Res. Group Wood Preserv., 1989, Doc. No. IRG/WP/3542.

6. D. J. Dickinson and R. J. Murphy, Rec. Annu. Conv. Br. Wood Preserv. Assoc., 1989, 2.

7. Proc. - Annu. Meet. Am. Wood-Preserv. Assoc., 1990, 86, 248.

8. American Wood-Preservers' Association, "Book of Standards 1990 Standard - A3-89", American Wood-Preservers' Association, Stevensville, MD, 1990.

9. U.S. Borax & Chemical Corporation, "TIM-BOR[R] for Wood Preservation", U.S. Borax & Chemical Corporation, Los Angeles, CA, 1986.

10. B. S. W. Dawson, G. F. Parker, F. J. Cowan, M. C. Croucher, S. O. Hone, and N. H. O. Cummins, Anal. Chim. Acta, 1990, 236, 423.

11. J. T. Hatcher and L. V. Wilcox, Anal. Chem., 1950, 22, 567.

12. M K. John, H. H. Chuah, and J. H. Neufeld, Anal. Lett., 1975, 8, 559.

13. U.S. Borax Research Corporation, Unpublished Results, Anaheim, CA.

14. Submitted to American Wood-Preservers' Association Technical Meeting, Santa Fe, New Mexico, Subcommittee P-5, September 1990.

15. American Wood-Preservers' Association, "Book of Standards 1990 Standard - M3-81", American Wood-Preservers' Association, Stevensville, MD, 1990.

16. American Wood Preservers Bureau, "Quality Control and Inspection Procedures", American Wood Preservers Bureau, Lorton, VA, 1988.

17. American Wood-Preservers' Association, "Book of Standards 1990 Standard - M2-89", American Wood-Preservers' Association, Stevensville, MD, 1990.

18. American Wood-Preservers' Association, "Book of Standards 1990 Standard - C2-89", American Wood-Preservers' Association, Stevensville, MD, 1990.

19. S. T. Lebow and J. J. Morrell, Forest Prod. J., 1989, 39, 67.

20. D. R. Carr, Proc. World For. Congr., 1962, 3, 1552.

21. K. M. Harrow, New Zealand J. Sci. Tech., 1951, 32B, 33.

22. A. J. McQuire, New Zealand Wood Industries, 1974, July.

23. D. C. Pribich, Proc. - Annu. Meet. Am. Wood-Preserv. Assoc., 1989, 85, 87.

24. J. W. Roff, "Information Report VP-X-125", Forintek Canada Corp., Vancouver, Canada, 1984.

25. U. K. Department of the Environment, Building Research Establishment, "Technical Note No. 41", Prince Risborough Laboratory, England, 1969.

26. New Zealand Timber Preservation Council "Specification of the Minimum Requirements of the N. Z. Timber Preservation Council Inc. MP 3640:1988", Standards Association of New Zealand, Wellington, NZ, 1988.

27. British Wood Preserving and Damp-proofing Association, "British Wood Preserving Association Manual", British Wood Preserving and Damp-proofing Association, London, 1986.

28. Standards & Industrial Research Institute of Malaysia "Malaysian Standard MS 995:1986", Standards & Industrial Research Institute of Malaysia, Selangor, Malaysia, 1986.

29. Private Communication, Mr. Jack Norton, Queensland Forest Service, Indooroopilly, Queensland, Australia.

30. F. W. Brooks, This symposium.

31. R. D. Arsenault, Proc. - Annu. Meet. Am. Wood-Preserv. Assoc., 1988, 84, 8.

TBTO and TBTN — Safe and Effective Biocides for Wood Preservation

H. Schweinfurth, D. Ventur,

INDUSTRIAL CHEMICALS, SCHERING AG, PO BOX 1540, D-4709 BERGKAMEN, FEDERAL REPUBLIC OF GERMANY

and R. Länge

EXPERIMENTAL TOXICOLOGY, SCHERING AG, PO BOX 65 03 11, D-1000 BERLIN 65, FEDERAL REPUBLIC OF GERMANY

1. INTRODUCTION

The biocidal activity of triorganotin compounds has been known since the early 1950s.[1] Starting from the 1970s trialkyltin compounds were commercially used on a large scale as biocides in the fields of plant protection, antifouling, material protection and wood preservation.[2,3]

For the latter application particularly two trialkyltin compounds - tri-n-butyltin oxide (TBTO), with the synonyms bis-(tri-n-butyltin)oxide or hexa-n-butyldistannoxane, and its long-chain carboxylic acid ester tri-n-butyltin naphthenate (TBTN) - are used nowadays as fungicides worldwide.

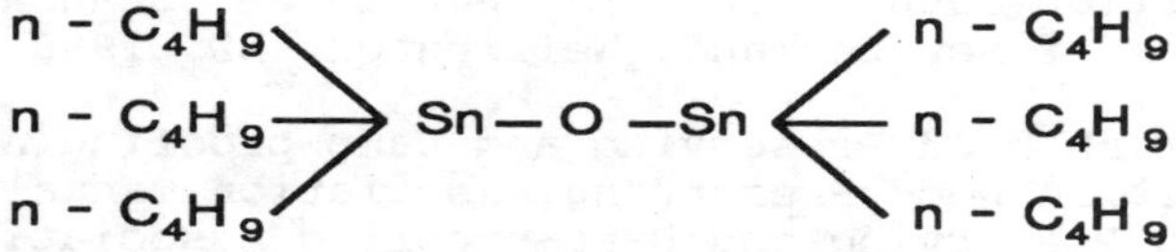

$$(n\text{-}C_4H_9)_3Sn-O-Sn(n\text{-}C_4H_9)_3$$

Tri-n-butyltin oxide (TBTO) (1)

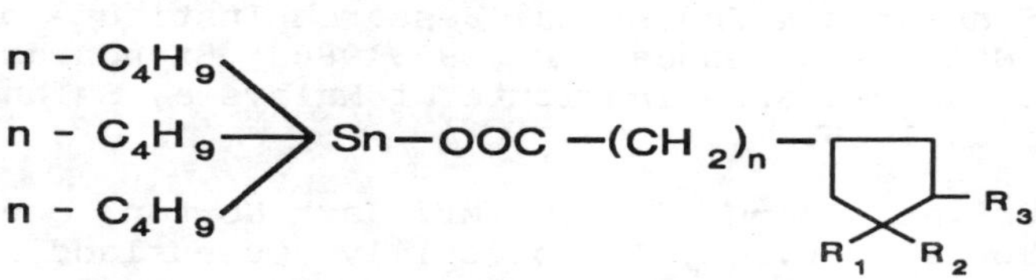

$$(n\text{-}C_4H_9)_3Sn-OOC-(CH_2)_n-\text{[cyclopentyl ring with } R_1, R_2, R_3\text{]}$$

Tri-n-butyltin naphthenate (TBTN) (2)

The following paper presents a survey of both technical and safety data on tributyltin oxide and tributyltin naphthenate. We will discuss the technical application of these two organotin compounds and their efficacy as wood preservatives.

The safety evaluation will focus on the results from comprehensive mammalian toxicity studies. Further consideration will be given to ecotoxicology and environmental fate data. Finally the status of registrations obtained for wood preservatives containing TBTO and TBTN will be described.

2. PHYSICAL AND CHEMICAL PROPERTIES

Table 1 summarizes some physical data of technical grade TBTO and TBTN.[4,5,6]

Table 1 Physical data of TBTO and TBTN

	TBTO	TBTN
Characteristics	colourless liquid	yellowish brown liquid
Density	1.17 g cm^{-3}	1.09 g cm^{-3}
Viscosity (20°C)	9 mPa*s	2900 mPa*s
Boiling point	173°C (1.3 hPa)	125°C (0.5 hPa)
Decomposition	> 230°C	> 230°C
Vapour pressure (20°C)	$1*10^{-3}$ Pa	$9*10^{-5}$ Pa
Flash point (DIN 51758)	190°C	78°C
Water solubility (20°C) (OECD Guideline No. 105)	71.2 mg/l	1.5 mg/l
Tin content	38.6 %	21 %

TBTN is formed by the reaction between TBTO and naphthenic acid. The latter "compound" is a mixture of different long-chain carboxylic acids.[7] Therefore the physical characteristics of TBTN may vary and are dependent on the type of naphthenic acid used for the production.

Both TBTO and TBTN have low volatility and water solubility so that they are not easily leached out of treated wood.[2,8,9,10] As seen from table 1 TBTN is still less volatile and less soluble than TBTO.

No restrictions or interferences with regard to the later working up of the wood were observed after fungicidal treatment with TBTO or TBTN. Furthermore the organotin compounds are compatible with many wood preservative ingredients, e.g. solvents, resins, pigments, fungicides and insecticides.

Concerning the chemistry of TBTO and TBTN in wood neither their chemical nature within the cellulose matrix nor their protection mechanism against attacks of fungi are well known. Different possibilities e.g. esterification of TBTO with terminal hydroxyl groups of cellulose or formation of carbonate species are under discussion.[3,11,12]

Some recent research has indicated that degradation of TBTO or TBTN within treated wood to lower butyltin species may take place under certain conditions.[13,14,15,16] Different factors were discussed for this phenomenon and stabilizers were proposed to improve the stability of organotin fungicides.[17,18]

Considering the 25 years of commercial use of TBTO in wood preservatives and the excellent performance in practice these results are not clearly understood to date and their relevance remains doubtful.[2,19,20,21]

3. EFFICACY DATA

Tri-n-butyltin oxide and tri-n-butyltin naphthenate have a broad range of fungicidal activity. They are most effective against wood destroying fungi (basidiomycetes), in particular against brown rot fungi which destroy the wooden cellulose.[2,11,12,19,22,23]

The good efficacy of TBTO-based wood preservatives against dry rot fungi was also reported.[24]

TBTO and TBTN are less active against white rot and soft rot fungi which attack both cellulose and lignin.

Compared to the concentrations used for brown rot fungi either higher concentrations of TBTO and TBTN or the combination with other co-biocides are consequently necessary for the control of these types of fungi.[9,25]
In general the efficacy of tributyltin compounds depends directly on the tributyltin content and is independent of the nature of the anion attached to the tri-n-butyltin group. Compared with TBTO, about double the quantity of TBTN is required to achieve the same fungicidal effectiveness.[11,26]

In the meantime a number of publications were issued about the toxic values of organotin compounds using different test methods.[2,3,19,25] The following table shows a comparison of toxic values for TBTO and TBTN against some brown rot fungi determined by the same methods.

Table 2 Toxic values for tri-n-butyltin oxide (TBTO) and tri-n-butyltin naphthenate (TBTN) against some fungi[5,27]

	TBTO	TBTN
	kg a.i./m^3	
Test fungus		
Coniophora puteana		
unaged	0.34 - 0.70	1.3 - 2.0
leached	0.68 - 2.11	1.3 - 2.0
evaporative aged	-	1.4 - 2.0
Poria monticola		
unaged	0.002 - 0.055	< 0.3
leached	0.021 - 0.055	< 0.3
evaporative aged	-	0.22 - 0.35
Lentinus lepideus		
unaged	0.054 - 0.144	< 0.3
leached	0.056 - 0.140	0.3 - 0.5
evaporative aged	-	0.55 - 0.88

Test methods: DIN 52176, EN 84, EN 73

Combination Products

As mentioned before, the toxic limits for white and soft rots are by far higher than those for brown rots. Moreover TBTO and TBTN are only moderately effective against blue stain and surface fungi which do not decay wood but can cause damage to decorative coatings. Furthermore the insecticidal activities of TBTO and TBTN are limited (see 3.3). To overcome these problems combinations of different specific fungicides and insecticides are formulated and used in wood preservation.

In fact such mixtures are advantageous as they give a broader spectrum of fungicidal and insecticidal activities than the individual components. In addition synergistic effects were observed in some cases.[28,29] The formulation shown in table 3 is an example for such a 'combination product' which exhibits a very broad spectrum of efficacy.

Table 3 Toxic values for a TBTN / Dichlofluanid / Permethrin formulation against fungi[30]

Formulation:	TBTN/Dichlofluanid/Permethrin
ratio a.i.	3 / 0.9 / 0.2
Test fungus	kg a.i./m^3
Coniophora puteana	
unaged	0.33 - 0.54
leached	0.75 - 1.22
evaporative aged	1.17 - 1.77
Poria monticola	
unaged	< 0.14
leached	< 0.20
evaporative aged	< 0.18
Gloeophyllum trabeum	
unaged	0.14 - 0.21
leached	< 0.20
evaporative aged	0.19 - 0.30
Blue stain test	
Aureobasidium pullulans	passed
Sclerophoma pityophila	
Larvae test	
Hylotrupes bajulus	passed
Test methods: EN 113, EN 84, EN 73, EN 152, EN 46 + EN 73	

Water-Based Wood Preservatives

The use of TBTO and TBTN as water-based wood preservatives has been limited due to their low water solubility (s. table 1). However, TBTO and TBTN can be emulsified in water using special emulsifiers. Also by the addition of suitable quaternary ammonium salts they become water-dispersible.[2,3]

Aqueous formulations based on TBTO or TBTN normally show an enhanced activity against wood destroying fungi compared to solvent-based formulations.[9,31,32] In addition water-based organotin wood preservatives can be formulated with other co-biocides resulting in a broader fungicidal activity and in an increased efficacy.

The good efficacy of an aqueous combination product based on TBTN and azaconazole [1-(2-(2,4-dichlorophenyl)-1,3-dioxolane-2-yl)-methyl-1H-1,2,4-triazole] is shown in Table 4.[33,34]

Table 4 Toxic values for an aqueous combination product against fungi[33,34]

	TBTN/Azaconazole
Formulation: ratio a.i.	2 / 1
Test fungus	kg a.i./m^3
Coniophora puteana	
unaged	0.25 - 0.40
leached	0.41 - 0.68
evaporative aged	0.66 - 1.06
Poria monticola	
unaged	< 0.24
leached	< 0.26
evaporative aged	< 0.28
Lentinus lepideus	
unaged	< 0.25
leached	0.27 - 0.40
evaporative aged	0.66 - 1.05
Coriolus versicolor	
unaged	1.30 - 2.30
leached	0.63 - 0.89
evaporative aged	0.66 - 1.09

Test methods: EN 113, EN 84, EN 73

From the data in table 4 it is apparent that the aqueous combination product attains lower toxic values against C. puteana than the organotin compounds alone in solvent-based systems (see table 2).

Even a good efficacy against the organotin less susceptible white rot fungi e.g. Coriolus versicolor could be achieved by the combination of TBTN with azaconazole.

This efficacy is mainly caused by synergistic effects on the biological activities of the individual components.

Insecticidal Activity

Both organotin compounds TBTO and TBTN are toxic to certain species of insects. But at concentration levels normally employed in wood preservation i.e. 1-3% their insecticidal effectiveness is limited.[11,25]

Therefore they are often combined with other organic contact insecticides, particularly chlorinated hydrocarbons. In some cases these combinations show synergistic effects, i.e. an increase concerning the fungicidal and insecticidal activities of the individual components.[35]

Some of the new synthetic pyrethroid insecticides which are less hazardous to the environment than the organochlorines are unstable in the presence of TBTO.[36,37,38] This incompatibility is due to some kind of chemical reaction and is restricted to TBTO only.

Tri-n-butyltin naphthenate, on the other hand shows no such reactivity and is totally compatible with synthetic pyrethroids.

4. APPLICATION METHODS

Due to the low aqueous solubility of tri-n-butyltin oxide and -naphthenate they are industrially applied to timber in organic solvents. The concentrations normally employed are between 1% and 3%.

For the treatment processes different methods such as vacuum and pressure impregnation or non-pressure methods such as dipping, spraying or in some cases brushing or deluging techniques are used.

The most broadly utilized methods for TBTO and TBTN are the pressure and the vacuum processes. With regard to environmental contamination and the effectiveness these closed-system treatments are preferable compared to the non-pressure application methods. Furthermore non-pressure methods usually provide a lower degree of protection because of the smaller amount of absorbed fungicide.

Water-soluble organotin formulations or emulsified solutions of TBTO or TBTN are mainly applied by pressure or immersion methods.

Apart from the application methods the degree of the fungicidal protection further relates to the type and state of timber and the solvent used in the process. In general TBTO and TBTN show a very good processing performance i.e. good distribution within and deep penetration into the wood.

5. STUDIES ON SAFETY FOR HUMANS

Toxicokinetics

In studies using ^{113}Sn-labelled TBTO and TBTN the gastrointestinal absorption in rats was found to be incomplete (approx. 30-55%) and slow.[39,40,41] Peak plasma levels occurred within 12-24 h after oral dosing. Whole body autoradiography investigations of the distribution of ^{113}Sn-label into tissues and organs of rats showed that both compounds were mainly localized in the excretory and metabolizing organs, i.e. liver and kidney, and - due to incomplete absorption - in the whole gastrointestinal tract . No specific enrichment in any tissue was noted for either compound.[39,40]

In the baboon monkey, about 10-15% of a dose of undiluted TBTO was absorbed through the intact skin.[39]

After oral administration of ^{113}Sn-TBTO and -TBTN to rats by gavage, the excretion occurred mainly via the bile, biphasically with half-lives of about 12-13 hours and 3 days.[39,40,41] When mice were given ^{14}C-TBTO in the drinking water for 5-30 days, they excreted the radioactivity mainly in the faeces (little absorption).[42] Kidney, liver, spleen and fat tissue contained the highest ^{14}C-activity. After termination of treatment the excretion was relatively rapid.

Hydroxylation of tributyltin compounds takes place mostly in the liver with subsequent dealkylation to di- and monobutyltin compounds and finally inorganic tin.[43,44] Separation of TBTO and its metabolites revealed that mostly unchanged compound was present in the lipophilic tissues of the rat (fat tissue, carcass) and that liver and kidney contained mainly metabolites.[39] In radiochromatograms of urine from rats that were dosed with ^{113}Sn-TBTO or -TBTN, respectively, the parent compounds themselves were almost undetectable.[39,40]

A moderate accumulation of TBTO and/or its metabolites was observed in the rat after repeated oral administration.[39] It was estimated that steady state conditions would be reached after about 3-4 weeks.

Acute Mammalian Toxicity

The results of acute toxicity studies with TBTO and TBTN in mammals are summarized in table 5.

Oral Uptake. The oral LD_{50} in rats was found to be 127 mg/kg for TBTO and 224 mg/kg for TBTN.[45] If the LD_{50} is expressed on the basis of the TBT content of the two compounds (i.e. in mmol/kg body weight), a rather good agreement is apparent.

Observations after treatment also were similar for both compounds. The predominant clinical signs were apathy and emaciation. Death occurred between days 2 and 6 after administration, i.e. with some delay. At necropsy of animals that had died, mainly inflammatory reactions within the gastrointestinal tract were noted. The LD_{50} values in rats for TBTO formulations in a range of concentrations that may be of interest, are assembled in table 6.[52] The acute oral toxicity showed a clear dependence on the TBTO content, but was low in all cases.

Table 5 Acute mammalian toxicity data[45-50]

Route	Species	Parameter	TBTO	TBTN
oral	rat	LD_{50} [mg/kg]	127	224
	mouse	LD_{50} [mg/kg]	152/92	---
dermal	rabbit	LD_{50} [mg/kg]	11.700	---
	rat	LD_{50} [mg/kg]	605	4.600
inhalative (aerosol)	rat	LC_{50} [mg/m^3]	65	152
inhalative (vapour)	rat	mortality (7h)	none	none

Table 6 Oral LD_{50} in rats for formulations containing TBTO[52]

Concentration TBTO [%]	Vehicle	LD_{50} [g/kg]
0.9	White spirit + additives	>4.0
1.3	White spirit + additives	8.0
2.5	Water + benzal-konium choride	3.9
5.0	White spirit	2.0

Dermal Contact. Upon dermal application the LD_{50} of TBTO varies between 605 mg/kg in the rat and 11,700 mg/kg in the rabbit.[47,48] This discrepancy can be attributed to differences in the formulations tested rather than to differences in the sensitivity of the animal species. For undiluted TBTN an LD_{50} of 4,600 mg/kg in rats was reported.[49] These results indicate that the acute toxicity of TBT compounds is considerably lower upon dermal contact when compared to oral intake.

Inhalative Uptake. In inhalation studies using rats both TBTO and TBTN showed a marked toxicity when sprayed.[45,50] The LC50 after 4 hours exposure was determined as 65 mg/m^3 for TBTO and 152 mg/m^3 for TBTN considering only inhalable particles of less than 10 µm diameter. Moreover, after a 1h exposure of guinea-pigs to TBTO at concentrations of 200 mg/m^3 and above (aerosol of a solution in olive oil) deaths were noted.[51]

On the other hand, rats were exposed to atmospheres saturated as possible with vapours of tributyltin compounds such as TBTO, TBTN or TBT benzoate once for 7 hours without mortality during exposure or the 14-days observation period.[45] Only minor clinical signs such as a slight nasal discharge directly after exposure were noted. These results indicate that vapours of the respective tributyltins are unlikely to cause an acute inhalation hazard.

Local Effects and Dermal Sensitization

Skin Irritation. The undiluted technical active ingredient TBTO was found to be severely irritating to the skin of rabbits and humans.[53,54] A comparative study on the skin tolerance of model wood preservatives which contained 1% TBTO, 2% TBTN or 2% TBT linoleate in a white spirit vehicle with 10% alkyd resin was carried out in human volunteers.[55] After a 30 min. exposure, all formulations tested were tolerated without significant reaction, whereas after 2h and 4h almost all application areas showed severe reddening and slight (occasionally moderate) swelling. There was, however, no remarkable difference between the effects of the preparations containing TBT compounds and those of the vehicle. In each case, the irritation after a 4h exposure was classified as slight to moderate. Thus it can be concluded that the observed local irritation was primarily attributable to the vehicle.

Eye Irritation. In the rabbit eye TBTO, TBTF (both compounds applied as active ingredients or in paint formulations) as well as TBTCl produced severe irritation.[53,56] Certain aqueous preparations containing 0.15 - 2.0% TBTO caused severe damage in the rabbit eye.[57]

Skin Sensitization. No evidence of allergenic effects on the skin of TBTO and TBTN was found in the maximization test in guinea pigs.[46,58] (Because of the locally irritating potential of the TBT compounds only low concentrations could be applied.) These results were corroborated by a study of two TBTO-based antifouling paints in man.[59]

Toxicity after Repeated Dosing

In toxicity testing with multiple doses information on the degree of exposure likely to cause harmful effects upon repeated exposure is obtained. Dose levels which are tolerated by experimental animals without adverse reaction are designated as "no-observed-effect level". In the planning of such experiments TBTO has been selected by us as a model substance for all the tributyltin derivatives currently of technical interest. Reasons for this are the good agreement among toxicokinetic as well as acute toxicity data for these compounds. In addition, there is experimental evidence that in aqueous media tributyltin esters as well as TBTO undergo dissociation, so that in each case the same tributyltin species is most probably involved in interactions with biological material.[60]

Repeated-dose Oral Toxicity. The systemic toxicity of TBT compounds after oral uptake has been studied extensively, in particular in rodents. Detailed reviews were published recently.[61,62,63] This paper gives a summary of the most prominent effects, the target organs and the no-effect levels found in different studies.

When TBTO was given to rats at dietary concentrations of 500 ppm over 4 weeks or at an oral dose of 25 mg/kg over 10 days, high mortality rates were noted.[45,64] No deaths occurred at 100 ppm in the diet or after daily oral administration of 12 mg/kg over 90 days each.[45,65]

General signs of toxicity, such as a reduced intake of food and a decrease in weight gain, were observed in rats receiving TBTO at 50 ppm in the diet over 4 weeks and up to 2 years.[66,67] Such effects occurred also in mice at dietary levels of 150 ppm TBTCl or 200 ppm TBTO.[68,69]

Toxic damage on the liver after administration of TBTO has been reported in three mammalian species. Necrosis of the liver and inflammatory reactions of the bile duct were found in feeding studies at 320 ppm in rats (4 weeks) and at 80 ppm in mice (13 weeks).[70,69] In dogs slight histological changes of the liver (vacuolization of periportal hepatocytes) were noted after an oral dose of 10 mg/kg over 8 to 9 weeks.[71] These changes were generally accompanied by increased liver weights and increased serum activities of liver enzymes.

Some hematological changes were induced in rats fed TBTO at 80 ppm over 4 weeks.[70] These consisted of reductions in hemoglobin concentration and erythrocyte volume and decreased lymphocyte counts. A decrease in red blood cell count (together with reduced hematocrit and hemoglobin values) was noted in mice which had received TBTO at 80 ppm in their diet over 13 weeks.[69]

Histological changes indicating a decreased activity of the pituitary-thyroid axis were noted in rats in short- term and long-term feeding studies with TBTO at 80 ppm or 50 ppm, respectively.[70,67] These were accompanied by changes in serum hormone concentrations in the short-term experiment only.

The formation of erythrocyte rosettes in the mesenteric lymph nodes of rats has also been reported from 4-week feeding studies with TBT compounds.[70,72,73] This change also arises spontaneously in this species, but could not be observed in a long-term feeding study using

TBTO.[67] The mechanism of its development and its biological significance are still unclear.

It appears, however, that the main targets of the toxicity of TBT compounds are the organs of the lymphatic system. The thymus weight was reduced in juvenile rats which had received TBTO in the diet at 50 ppm for 4 weeks or at 100 ppm for 13 weeks.[66,45] Other authors found the same effect after 4 weeks of treatment with as little as 20 ppm TBTO (only in male animals).[70] A reduction in rat thymus weight was also observed after short-term feeding of TBTN at 40 ppm and of TBTCl at 100 ppm.[73,72] A dose-range finding experiment in dogs indicated that thymus atrophy also occurs in this species at 10 mg/kg TBTO over 8-9 weeks.[71] In comparative studies in which TBTCl was administered to rats, mice and Japanese quail for 2 weeks in the diet, it was demonstrated that rats are considerable more sensitive than the two other species with respect to thymus atrophy.[68]

Short-term tests on immune function in rats revealed a suppression of thymus-dependent immune responses and of nonspecific resistance at TBTO concentrations of 20 ppm and more in the diet.[74] Results from long-term feeding studies showed that TBTO at a dietary level of 50 ppm suppressed the resistance to bacterial (Listeria monocytogenes) and to parasitic (Trichinella spiralis) infection both in young and in aged rats.[75] Additionally impairment of resistance to the nematode T. spiralis was reported in young rats at 5 ppm TBTO. The relevance of the latter finding is unclear. Further 4-week studies in juvenile rats revealed a no-effect level of 5 ppm (equivalent to 0.6 mg/kg body weight) with regard to alterations of immune competence.[66]

In conclusion from the various studies reviewed in this chapter, the no-effect levels (NEL) found in feeding experiments with TBTO and TBTN are summarized in table 7. Both compounds have very similar NEL's in subacute studies in rats when expressed in terms of their TBT content. This again demonstrates their close similarity with regard to biological effects.

Repeated-exposure Inhalation Toxicity. In an inhalation study with TBTO, juvenile rats were exposed under "nose only" conditions 4 hours daily, apart from weekends, for 29-32 days.[45] With a total concentration of

<u>Table 7</u> No-adverse effect levels (NEL) in feeding studies

Compound	Species	Duration	NEL	Ref.
TBTO	rat	28 d	5 ppm	66
		90 d	4 ppm	45
		2 yrs.	5 ppm	67
			(0.5 ppm)*	75
	mouse	90 d	20 ppm	69
TBTN	rat	28 d	8 ppm	73

*NEL after infection with trichinella spiralis (see text)

2.8 mg/m^3 TBTO as aerosol there were marked toxic effects which included mortality. Histological investigation revealed inflammatory reactions in the whole of the respiratory tract and changes in the lymphatic organs like those occurring after oral uptake of large doses.

After inhalation of TBTO vapour (0.03 or 0.16 mg/m^3) neither local nor systemic toxic effects were observed. The concentration of 0.16 mg/m^3 TBTO corresponding approximately to the equilibrium vapour pressure at room temperature can thus be taken as a no-observed-effect level for the rat.

<u>Neurotoxicity</u> The short-chain homologues of the tributyltin compounds are known as neurotoxic agents. Trimethyltin compounds cause neuronal damage in certain regions of the brain, whereas triethyltin compounds induce oedema of the white substance of the brain and spinal cord.[76,77] In the older literature there are also indications that tributyltin compounds could possibly damage the central nervous system.[56,78] In more recent studies on the effects of TBTO, TBTCl and TBTN, no morphological changes of the brain and spinal cord were found even though this aspect was given particular attention in some studies. Similarly, none of the behavioural changes typical of the effect of the lower homologues were observed. Therefore a neurotoxic potential of TBT compounds is very unlikely.

Mutagenicity

TBTO has been tested comprehensively for mutagenic properties in short-term tests in vivo and in vitro. Such studies were conducted within a multi-centre trial initiated by the World Health Organization (WHO) as well as in our laboratories.[79,80]

TBTO caused neither gene mutations in Salmonella typhimurium in the Ames test nor genetic changes (gene conversions) in the yeast Saccharomyces cerevisiae.[79,80] However, in the WHO study an increase in the number of revertants which was not dose-dependent occurred in the fluctuation test in one strain of S. typhimurium (TA 100) in the presence of rat liver S9 mix.[79] This finding is of questionable biological significance. No evidence of point mutations or DNA damage was found in other in vitro test systems with microbial or mammalian cells.[79]

In an experiment in vitro using Chinese hamster ovary cells, chromosome damage was observed at the highest concentration in the presence of S9 mix.[79] A connection with the cytotoxicity which was also seen in this study may be surmised. In addition, the same authors reported a positive result in the micronucleus test in male mice 48 hours after administering a single oral dose of 60 mg/kg. A subsequent re-evaluation of the bone marrow preparations from this WHO study produced no evidence of a mutagenic potential.[81] In our laboratory no evidence of chromosome mutations was found with TBTO, either in human lymphocytes in vitro or in the micronucleus test in the mouse, although the tested concentrations and doses extended into the distinctly cytotoxic range.[80] In the last-mentioned micronucleus test 5000 polychromatic erythrocytes were scored instead of 1000 cells. The highest administered dose, 125 mg/kg p.o., was about twice as high as that in the WHO study.

The results of the vast majority of tests with procaryotic or eucaryotic cells indicate that any potential of TBTO to induce point mutations may be disregarded. If one takes into account the critical remarks about the positive results found sporadically as well as the convincingly negative result of the micronucleus test with a higher dose range and larger sample size, then the available data are also not indicative of a clastogenic potential of TBTO. Accordingly TBTO is not considered to have a mutagenic potential.[61,62]

Six tributyltin "esters" including TBTN also did not induce point mutations in the Ames test. Furthermore TBTN was shown to be negative in the mouse micronucleus test after oral administration of up to 500 mg/kg.[80]

Carcinogenicity

In a two-year study on chronic toxicity and carcinogenicity TBTO was given to groups of 50 male and 50 female rats in the diet.[67] The concentrations were 0, 0.5, 5.0 and 50 ppm. (Results on chronic toxicity were reviewed above.) The adverse effects on body weight gain in the second year indicate that 50 ppm was the maximum tolerated dose. In this high dose group an increased incidence of benign pituitary tumours (prolactinoma) and of adrenal medullary tumours (phaechromocytoma) was observed in both sexes. Furthermore the number of parathyroid adenomas was increased only in male animals of the 50 ppm group whereas that of tumours of the adrenal cortex was reduced.

Based on the above-mentioned mutagenicity studies it is unlikely that TBTO has a potential to interact with the genetic material. Prolactinomas and phaechromocytomas had been observed spontaneously in the rat strain used at a high and variable incidence. Tumours of the adrenal medulla are often found in rats together with multiple endocrine tumours. Therefore, the above-mentioned increase in the incidence of tumours is considered an epigenetic effect, probably a result of the described hormonal or immunotoxic effects of TBTO. Unless humans are exposed to similarly high doses for extended periods, an increased tumour risk is not to be expected.[61]

Developmental and Reproductive Toxicity

TBTO was studied for embryotoxic including teratogenic effects in mice, rats and rabbits.

When TBTO was administered to mice orally by gavage on days 6 through 15 of pregnancy, no adverse effects on the dams or fetuses were noted up to the dose of 5.8 mg/kg.[79] An increased frequency of cleft palates was observed at doses of 11.7 mg/kg and higher. These doses also produced maternal toxicity (reduced body weight) which may explain the occurrence of cleft palates, since it is known that this abnormality can be non-specifically induced in mice (e.g. by stress). Additionally the two highest doses of 23.4 mg/kg and 35 mg/kg caused a reduction of fetal weights and an increase in minor skeletal abnormalities and variations. At a dose level of 35 mg/kg prenatal mortality was increased.

In a further embryotoxicity study in mice no adverse effects were noted at 5 and 20 mg/kg TBTO (on days 6 to 15 of gestation).[82] The highest dose of 40 mg/kg produced clear signs of maternal toxicity (body weight depression) and embryotoxicity (increased resorption and decreased fetal weights). No malformations were reported, but it is unclear whether the methodology of fetal examination was suitable for detection of cleft palates.

After oral dosing of rats with TBTO on gestation days 6 through 19, no adverse effects on the mothers or the embryos were noted at the low dose of 5 mg/kg.[83] Slightly decreased body weight gain (indicating a marginal maternal toxic effect) and a slight retardation of ossification in the fetuses were observed at 9 mg/kg. Slight embryolethal effects as well as a marked retardation of fetal development and teratogenic effects (mainly cleft palates) were noted only after the highest dose of 18 mg/kg at which maternal body weight gain was markedly decreased. (These results are in general accordance with those in a prenatal toxicity study in which rats were dosed intragastrically with TBTO on days 6 to 20 of gestation and allowed to litter spontaneously.[84])

The teratogenic potential of TBTO in rabbits was studied after oral administration on gestation days 6 through 18.[85] The no-effect level for maternal toxicity and embryotoxicity/fetotoxicity was 1.0 mg/kg. At the highest dose of 2.5 mg/kg maternal toxic effects (body weight loss) as well as an increased incidence of abortions and a marginal retardation of fetal development, but no teratogenic effects were reported.

It can be concluded that TBTO had no embryotoxic or fetotoxic effects in three mammalian species at dose levels which were not toxic to the dams. Therefore toxic effects on the embryo or fetus (in particular teratogenic effects) need not be expected if exposure limit values will be observed that are suitable to protect adults from adverse effects.[61]

Furthermore the effects TBTO on reproductive performance and fertility were investigated in a study through two consecutive generations (F_0,F_1).[86] At a dietary level of 50 ppm signs of slight general toxicity were noted in the parental generations, as indicated by a reduced food consumption during the pre-mating period in F_0 females and lower body weights in F_1 animals. In the pups reduced weights were noted in both generations. Moreover, TBTO at 50 ppm caused a decrease in thymus weights in F_0 females and in F_1 males and females. However, in both generations there were no adverse effects on fertility or on pup survival at this dose. The no-observable-effect level for parental toxicity, fertility and developmental toxicity in this study was 5 ppm.

Effect on humans

The local effects of undiluted TBT compounds and of TBT-based model formulations for wood preservatives on human skin were already described above (see "Skin Irritation"). During the medical surveillance of organotin workers contact dermatitis of varying severity was observed in individuals who had accidental contact with TBT compounds[87]. The delayed character of the irritancy was found to lead to extended exposure in some cases. Prompt cleaning of the effected skin areas prevented the irritation.

Recently the results of yearly detailed medical examinations of the employees of an organotin manufacturer in the United States were published.[88] The only difference between workers from the inorganic and organotin production and new or unexposed employees were slightly, but significantly lower values for erythrocyte count, hemoglobin and hematocrit. However, all respective values remained within the normal range and when only data from tributyltin production workers are compared with those from office or maintenance staff, there are practically no differences in these red blood cell parameters.

Furthermore, the production workers at the plants of the major European organotin producers are subjected to yearly medical examinations which include a general physical examination as well as clinical blood chemistry and hematology studies.[87,89] In these examinations, but also in the special examinations performed after extensive accidental contamination, there have been no indications of adverse systemic effects of TBT compounds - even when the presence of organotin in urine samples was demonstrated.

6. CONCLUSIONS ON SAFETY FOR HUMANS

Although TBT compounds have been produced for more than 25 years and used on a fairly broad scale in wood-preservation and other applications, there are no reports on cases of acute systemic poisoning or on long-term adverse effects in workers.

Exposure of employees to wood preservatives containing TBTO or TBTN can occur during their production and use. Contact will mainly be via the dermal route or via inhalation. It is obvious that the high irritancy of the undiluted active ingredients and of formulations containing higher concentrations requires prevention of contact with the skin and the eyes (protective gloves, goggles or masks according to the instructions given by the manufacturers). Contaminated working clothes should be changed immediately. In case of aerosol formation, e.g. when it is necessary to spray tributyltin formulations, adequate respiratory protection is required to preclude inhalation of the aerosol.

Occasional accidental skin contact with end-use formulations containing 1% TBTO or 2% TBTN should not produce health hazards, because their irritancy as well as their dermal penetration and systemic dermal toxicity can be considered low. Furthermore elimination of tributyltins from the organism can be expected within a few days. Nevertheless adequate personal protective equipment (see above) should be worn to minimize exposure.

The following considerations apply to repeated exposure to TBTO vapour in the atmosphere at the workplace. In repeated-dose studies with oral administration of TBT compounds toxic effects on the liver and bile ducts as well as changes of blood cells and endocrine organs were observed at higher dose levels. Toxic effects on the lymphatic organs and an impairment of the immune system, in particular in host resistance models, were noted at still lower doses in rats. However, no adverse effects were found in rats after repeated exposure (4 weeks) to an atmosphere that was practically saturated at room temperature with TBTO vapour at 0.16 mg/m^3. Therefore a provisional MAK value of 0.05 mg/m^3, as TBTO (vapour), was established for TBT compounds in Germany.[61] In other countries, such as the United Kingdom and the United States, these compounds are still covered by the general exposure limit value for organotin compounds of 0.1 mg/m^3 as tin.

Due to the strong fixation of organotin compounds in the wood and their low potential to leach or evaporate, there is no evidence for any health hazards by handling wood impregnated with tributyltins.

It was shown recently that no adverse effects have to be expected from long-term exposure of man to vapours in rooms with large surfaces of TBTN-treated wood.[45] Nevertheless such application in indoor areas is not generally recommended.

Finally, the results described above (chapter 5) indicate that mutagenic, carcinogenic or teratogenic effects of tributyltin compounds do not have to be expected at the workplace or in areas where TBT-treated wood has been used.

7. STUDIES ON ENVIRONMENTAL BEHAVIOUR

Wood preservatives containing tributyltin compounds may be introduced into the environment during accidental spill situations. Some treatment processes may also result in smaller emissions of TBT into natural ecosystems. Based on these potential routes of input, in the following the ecological fact and effects on fresh water and terrestric soil ecosystems are discussed, which are primarily affected by the above mentioned events.

Ecological Fate of Tributyltin Compounds

As summarized in previous reviews the important and biologically relevant removal pathway of tributyltin compounds is their debutylation leading eventually to the inorganic form.[90,91] Abiotic removal processes, like volatilization, hydrolysis or thermolysis, are negligible in respect of elimination of TBT from environmental media.

Degradation in Water. In experiments on the photolysis of tributyltin in freshwater the degradation half-life of TBT was approx. 25 hours at irradiation with light of 300 nm wave length, while at 350 nm the half-life was approx. 18 days. Under natural sunlight, the degradation half-life was estimated to be more than 89 days under the specific conditions during the experimental period (summer 1981). The presence of photosensitizing agents like fulvic acid, increased the photolytic degradation considerably.[92]

Similar findings were reported with half-lives of 18.5 days and 3.5, respectively, then irradiated with UV-light (wave length not reported) and in the absence or presence of acetone.[93]

The degradation of tributyltin compounds by microbial freshwater organisms was studied under various conditions (see table 8). In summary, the half-life for microbial aerobic degradation was reported to range from a few days up to 5 months. In one case, where total mineralization was analyzed, after 126 days, 13-20% of the TBT compounds were degraded. The differences can be explained by the varying test conditions, e.g. temperature and light regime, microbial populations and further test parameters. It can be concluded that tributyltins are biodegradable in freshwater systems, but that the rate of degradation is dependent on the particular environment.

Taking into account that natural sunlight reaches the water surface with wave lengths of 290 nm and above, with a more or less rapid attenuation in the water column depending on the amount and nature of suspended or dis-

Table 8 Biological degradation of tributyltin compounds

Test conditions	**Concentration (μg/l as TBT)**	**Half-life (weeks)**	**Ref.**
Freshwater (Toronto Harbour), shake flask, 20°C, dark incubation	0.025 - 0.5	20 ± 5	94
Freshwater/sediment mixture (Toronto Harbour) shake flask, 20°C, dark incubation	1 mg/kg sediment (dry weight)	16 ± 2	94
Microbial cultures, shake flask, 24-37°C	1000 - 2500	0.5 - 1.5	95
Algal culture, shake flask, 20°C	20	4	96

solved matter, the importance of photolytic degradation seems to be less significant than the biological degradation, perhaps with the exception of TBT associated with the surface microlayer.

Degradation in soil. The photolytic degradation in terrestric soil was investigated using soil-coated plates contaminated with TBT compounds. TBTO was degraded under irradiation with sunlight with a half-life of 21 days.[93] Half-lives of 7 to 10 weeks for the loss of parent compound were found for microbial degradation under aerobic conditions.[93] In a further experiment degradation half-lives of 15 to 20 weeks for soils were reported, yet it is not clear if conditions were aerobic all through the experimental period.[97]

Waste water treatment. Of particular interest is the degradation of tributyltin compounds in waste water treatment plants. In a laboratory waste water treatment system TBTO was degraded to more than 87% after an adaption period of more than 5 weeks.[98]

In Swiss sewage treatment plants about 98% removal of butyltin compounds was found.[99] The levels of butyltins observed in the sewage sludge indicated that elimination of tributyltin compounds occurred mainly by absorption to the sludge.

Ecological Effects

A considerable number of investigations have been published regarding the ecological effects of tributyltin compounds on various organisms. In respect of freshwater organisms, results are summarized in table 9.

It can be concluded that the short term exposure to TBT causes adverse effects in the most sensitive freshwater organisms in the range of 1 to 10 µg/l, whereas the long-term toxicity values are approximately one order of magnitude lower.

Table 9 Ecological effects of tributyltin (TBT) compounds in freshwater organisms

Test species	Effect/exposure period	Effective concentration TBT (μg/l)	Ref.
Microorganisms:			
Pseudomonas aeruginosa	MIC/72h	300,000	100
Mixed bacterial culture	IC50 (O_2 consumption)/2h	2,000	101
Bacillus cereus	MIC	500	102
Various species (algae, phytoplancton)	IC50/4h (primary productivity)	3-20	103
Molluscs:			
Lymnea stagnalis	LC50/not reported	42	104
Crustaceans:			
Daphnia magna	EC50/48h	10	105
	NOEC/21d	0.2	106
Artemia salina	LC50/48h	1,240	107
Fish:			
Rainbow trout	LC50/96h	3.5-7.3	108/109
	NOEC/120d	0.2	110
Lake trout	LC50/96h	13	109
Guppy	LC50/96h	21	104
	NOEC/not reported	0.32	104
Bluegill sunfish	LC50/96h	8.5	111

8. CONCLUSIONS ON SAFETY FOR THE ENVIRONMENT

Due to the usage pattern of TBT compounds in wood preservation the main aspect for consideration is the fate and effect in freshwater ecosystems and in terrestric soils. Considering the high aquatic toxicity of tributyltin compounds, measures must be taken to avoid any spillage into waters. Furthermore TBT-based preservatives should not be used for treatment of wooden structures that are immersed into water.

The fate and effect data, presented in chapter 7, led to the conclusion that aquatic life in freshwater is not adversely effected by TBT concentrations of 20 ng/l and below, if the organisms are continously exposed to TBT, following the regulations of the British Department of the Environment.[112] This environmental quality standard is based on the lowest toxicity value for freshwater species and includes a safety factor of 10. Employing the same procedure for short-term exposure, a safe level would be in the range of 300 ng/l. Carr et al. (1987) calculated a TBT concentration of 26 ng/l as an advisory concentration for freshwater ecosystems, using the guidelines developed in the United States for deriving water quality criteria. This value is in good agreement with the above-mentioned U.K. standard. To meet these standards, it is necessary to keep all discharges from treatment plants to the minimum level.

There are indications that biodegradation can occur in terrestric soil (probably slowly). Tributyltins will be adsorbed strongly, and therefore their bioavailability to soil organisms and the hazard of groundwater contamination should be low. Accordingly the general hazard of these compounds to the terrestrial environment is likely to be low.[62]

9. DISPOSAL OF WASTE

Spilled solutions should be collected by using absorbing materials such as sawdust or fine sand which is then disposed of in the manner described below.

If necessary TBTO or TBTN can be removed from polluted water by certain activated carbons as adsorption material.[114] Contaminated soil must be removed and either incinerated or taken to special refuse tips. Large amounts of organotin compounds or adsorption materials containing TBTO or TBTN can either be destroyed by burning in special incinerating plants or be deposited at special waste disposal sites according to local or government regulations.

Furthermore smaller amounts of TBTO and TBTN can be decomposed to inorganic tin under strong oxidation conditions e.g. heating with concentrated acids or treatment with potassium permanganate or hydrogen peroxide in acid solution.

10. STATUS OF APPROVAL

Wood preservatives containing TBTO and TBTN have found approval in a large number of European countries (see table 10). It should be noted that presently there is no uniform registration procedure even within the European Community. In the United Kingdom approval for wood preservatives containing TBTO and TBTN extends to industrial use only. The approval for TBTO formulations in household and professional use was withdrawn recently (with the exception of certain pastes). Some exposure data had raised concern that after remedial treatment of houses the safety factor might not always be as high as desirable. Obviously these considerations are applicable only for this particular application. In Germany wood preservatives containing tributyltins were evaluated for their safety by the Federal Health Office (BGA) and were registered in the schemes of IBT (industrial use) and RAL (do-it-yourself application).

In non-European countries TBTO-based wood preservatives have been registered in the United States, Canada and Australia. In Japan TBTO has been designated a "Class I specified chemical" recently and thus banned for all uses because of concerns that it may bioaccumulate in seafood as a consequence of its extended use in antifouling paints. Applications for approval of TBTN were submitted in a number of mainly Asean countries and are presently being evaluated.

Table 10 Status of approval in Europe

Country	TBTO	TBTN
Austria	yes	yes
Denmark	yes	yes
France	yes	yes
Germany	yes	yes
Netherlands	yes	no
Sweden	yes	yes
U.K.	industrial	industrial

REFERENCES

1. G.J.M. van der Kerk, J.G.A. Luijten, J.Appl.Chem., 1954, 4, 314; 1956, 6, 56; 1957, 7, 369; 1959, 9, 106; 1961, 11, 35
2. C.J. Evans, S. Karpel, 'Organotin Compounds in Modern Technology', J. Organomet. Chem. Lib. 16, Elsevier 1985
3. S.J. Blunden, P.A. Cusack, R. Hill, 'The Industrial Uses of Tin Chemicals', The Royal Society of Chemistry, London, 1985, Whitstable
4. Schering AG, Techn. Data Sheet 'Organotin Compounds'
5. Schering AG, Techn. Data Sheet 'Eurecid^R 9240 (TBTN)'
6. Schering AG, Techn. Data Sheet 'Eurecid^R 9000 (TBTO)'
7. E.S. Lower, Chem. Week, 1987, April, 76
8. H. Landsiedel, H. Plum, Holz als Roh- und Werkstoff, 1981, 39, 261
9. R. Hill, A.H. Chapman, B. Patel, A. Samuel, Int. Res. Gp. Wood Preserv., 1986, Doc. IRG/WP/3390
10. A. Gambetta, E. Orlandi, C.R. Cramer, D. Maier, H.P. Tscholl, Mater. und Organism., 1988, 23, 61
11. C.J. Evans, R. Hill, J. Oil Col. Chem. Assoc., 1981, 64, 215
12. B.A. Richardson, Proc. 1st Int. Biodet. Sym., 1968, 498
13. M.-L. Edlund, B. Henningsson, Int. Res. Gp. Wood Preserv., 1987, Doc. IRG/WP/3419
14. M.-L. Edlund, J. Jermer, B. Henningsson, W. Hintze, Int. Res. Gp. Wood Preserv., 1985, Doc. IRG/WP/3339
15. S.J. Blunden, R. Hill, ITRI Publication No. 698, 1988
16. S.J. Blunden, 31. Deutscher Zinntag, Düsseldorf, 1989
17. R. Hill, A.H. Chapman, A. Samuel, K. Manners, G. Morton, Int. Res. Gp. Wood Preserv., 1984, Doc. IRG/WP/3311
18. F. Imsgard, B. Jensen, H. Plum, H. Landsiedel, Brit. Wood Preserv. Assoc. Ann. Conv., 1985
19. C.J. Evans, Polym. Paint Col. J., 1990, 180, 428
20. H. Plum, H. Landsiedel, Holz als Roh- und Werkstoff, 1980, 38, 461
21. R. Hill, A.J. Killmeyer, Am. Wood Preserv. Assoc., 1988, 131

22. H. Greaves, M.A. Tighe, D.F. McCarthy, Int. J. Wood Preserv. , 1982, 2, 21
23. B.A. Richardson, 'Wood Preservation', The Construction Press, Lancaster, 1978
24. M. Takahashi, K. Nishimoto, Int. Res. Gp. Wood Preserv., 1985, Doc. IRG/WP/2238
25. H. Becker, Seifen-Öle-Fette-Wachse, 1987, 113, 773; 1988, 114, 27; 1988, 114, 61; 1988, 114, 99
26. A.F. Bravery, J.K. Carey, Int. Res. Gp. Wood Preserv., 1985, Doc. IRG/WP/3333
27. A. Bokranz, H. Plum, 'Technische Herstellung und Verwendung von Organozinnverbindungen', 1975, Schering AG, Industrie-Chemikalien, korr. und erweiterte Neuaufl. von: Fortschr. Chem. Forsch. 1971, 16 (3/4), 356
28. H.A.B. Landsiedel, Int. Res. Gp. Wood Preserv., 1987, Doc. IRG/WP/3426
29. Schering AG, Techn. Information 'Wood Protection', 1988
30. Testing results for Bayer AG/Schering AG according to the RAL-GZ 830 quality and test regulations, Bundesanstalt für Materialprüfung, Berlin, 1990
31. A.F. Bravery, Int. Biodet. Bull., 1970, 6(4), 145
32. T.R.G. Cox, Int. J. Wood Preserv., 1979, 1, 173
33. Schering AG, Techn. Data Sheet 'XE 9012'
34. Schering AG, Techn. Data Sheet 'TF 2052'
35. U. Thust, Tin and its Uses, 1979, 122, 3
36. S.J. Blunden, R. Hill, Int. Res. Gp. Wood Preserv., 1987, Doc. IRG/WP/3414
37. S.W. Carter, Rec. Ann. Conv. Brit. Wood Pres. Assoc., 1984, 132
38. J. Blunden, R. Hill, Appl. Organomet. Chem., 1988, 2, 56
39. M. Hümpel,G. Kühne, U. Täuber, P.E. Schulze, "ORTEPA-Workshop, Toxicology and Analytics of the Tributyltins -The Present Status, Berlin, May 15-16, 1986" ORTEP Association, Vlissingen-Oost, NL, 1987, p. 122
40. NATEC, Report on Project NA 879784, Hamburg, 1989 (Property of Schering AG)
41. M. Hildebrand, "Comparison of Toxicokinetics of TBTO and TBTN", Schering AG, Berlin, 1991
42. W.H. Evans, N.F. Cardarelli, D.J. Smith, J.Toxicol. Environm. Hlth., 1979, 5, 871
43. E.C. Kimmel, R.H. Fish, J.E. Casida, J. Agric. Food Chem., 1977, 25, 1

44. O. Wada, S. Manabe, "ORTEPA Workshop, Toxicology and Analytics of the Tributyltins - The Present Status, Berlin, May 15-16, 1986", ORTEP Association, Vlissingen- Oost, NL, 1987, p. 113
45. H. Schweinfurth, Tin and its Uses, 1985, 143, 9
46. P. Poitou, B. Marignac, C. Certin, D. Gradiski, Ann. Pharm. Franc., 1978, 36, 569
47. O.R. Klimmer, Arzneimittel-Forsch., 1969, 19, 934
48. J.R. Elsea, O.E. Paynter, Arch.Industr. Hlth, 1958, 18, 214
49. Scantox Biologisk Laboratorium, "Dermal LD_{50} determination of tributyltinnaphthenate", DK-Skensved, 1978 (Property of Gori all-wood, DK-Kolding)
50. H.Schweinfurth, Report IC 7/87, Schering AG, Berlin, 1987
51. R. Truhaut, Y. Chauvel, J.-P. Anger, N. PhuLich, J. van den Driessche, L.R. Guesnier, N.Morin, Europ. J. Toxicol., 1976, 9, 31
52. Schering AG, Experimental Toxicology, Internal Reports (ZK 21.955)
53. A.W. Sheldon, J. Paint. Technol, 1975, 47, 54
54. W.H. Lyle, Brit. J. Industr. Med., 1958, 15, 193
55. C. Schöbel, H. Wendt, "ORTEPA Workshop, Toxicology and Analytics of the Tributyltins - The Present Status, Berlin, May 15-16, 1986", ORTEP-Association, Vlissingen-Oost, NL, 1987, p. 180
56. R. Gohlke, E. Lewa, A. Strachovsky, R. Köhler, Z. Ges. Hyg., 1969, 15, 97
57. Z. Pelikan, Brit. J. Industr. Med., 1969, 26, 165
58. Scantox Biologisk Laboratorium, "Test for delayed contact hypersensitivity in the albino guinea-pig of tributyltinnaphthenate", Report, DK-Skensved, 1978 (Property of Gori all-wood, DK-Kolding)
59. M. Gammeltoft, Contact Dermatitis, 1978, 4, 238
60. W. Repenthin, "Allgemeine Physikochemie, Report 58a/86", Schering AG, Berlin, 1987
61. DFG, Deutsche Forschungsgemeinschaft, "Occupational Toxicants: critical data evaluation for MAK values and classification of carcinogens", VCH, Weinheim, 1991, p.315
62. World Health Organization, IPCS, "Tributyltin Compounds", Environmental Health Criteria 116, Geneva, 1990
63. H. Schweinfurth, P. Günzel, "Oceans '87 Proceedings, International Organotin Symposium, Halifax, Nova Scotia, Canada, Sept. 28-Oct.1, 1987", The IEEE Service Center, Piscataway, NJ, and The Marine Technology Society, 1987, Vol.4, p. 1421

64. H. Schweinfurth, "ORTEPA Workshop, Toxicology and Analytics of the Tributyltins - The Present Status, Berlin, May 15-16, 1986", ORTEP Association, Vlissingen-Oost, NL, 1987, p. 14
65. N. Funahashi, I. Iwasaki, G. Ide, Acta Path. Jap., 1980, 30, 955
66. F. Verdier, M. Virat, H. Schweinfurth, J. Descotes, Arch. Toxicol. Environ. Health (accepted for publication)
67. P.W. Wester, E.I. Krajnc, F.X.R. van Leeuwen, J.G. Loeber, C.A. van der Heijden, H.M.G. Vaessen, P.W. Hellemann, Fd. Chem. Toxic., 1990, 28, 179
68. N.J. Snoeij: "Triorganotin compounds in immunotoxicology and biochemistry", Thesis, Univ. of Utrecht, Utrecht, 1987, p. 55
69. Bio/dynamics Inc., Report on project. no. 87-3130, East Millstone, N.J., 1989, (Property of Aceto Corp., Atochem N.A. Inc. and Schering AG)
70. E.I. Krajnc, P.W. Wester, J.G. Loeber, F.X.R. Van Leeuwen, J.G. Vos, H.A.M.G. Vaessen, C.A. van der Heijden, Toxicol. appl. Pharmacol., 1984, 75, 363
71. H. Schweinfurth, Report IC 6/88, Schering AG, Berlin, 1988
72. N.J. Snoeij, A.J. van Iersel, A.H. Penninks, W. Seinen, Toxicol. appl. Pharmacol., 1985, 81, 274
73. H. Schweinfurth, Report IC 35/88 (ZK 22.688), Schering AG, Berlin, 1988
74. J.G. Vos, A. De Klerk, E.L. Krajnc, W. Kruizinga, B. van Ommen, J. Rozing, Toxicol. Appl. Pharmacol., 1984, 75, 387
75. J.G. Vos, A. De Klerk, E.I. Krajnc, H. van Loveren and J. Rozing, Toxicol. Appl. Pharmacol., 1990, 105, 144
76. A.W. Brown, R.D. Verschoyle, B.W. Street, W.N. Aldridge, J. Appl. Toxicol., 1984, 4, 12
77. I. Watanabe, in P.S. Spencer, H.H. Schaumburg (Eds.): "Experimental and Clinical Neurotoxicology", Williams and Wilkins, Baltimore, London, 1980, p. 545
78. J.M. Barnes, H.B. Stoner, Brit. J. Industr. Med., 1958, 15, 15
79. A. Davis, R. Barale, G. Brun, R. Forster, T. Günther, H. Hautefeuille, C.A. van der Heijden, A.G.A.C. Knaap, R. Krowke, T. Kuroki, N. Loprieno, C. Malaveille, H.J. Merker, M. Monaco, P. Mosesso, D. Neubert, H. Norppa, M. Sorsa, E. Vogel, C.E. Voogd, M. Umeda, H. Bartsch, Mutat. Res., 1987, 188, 65

80. R. Reimann, R. Lang, "ORTEPA Workshop, Toxicology and Analytics of the Tributyltins - The Present Status, Berlin, May 15-16, 1986", ORTEP Association, Vlissingen-Oost, NL, 1987, p. 66
81. R. Lang, Pharma Research Report IC 7/86, Schering AG, Berlin, 1986
82. S. Baroncelli, D. Karrer and P.G. Turilazzi, Toxicol. Lett., 1990, 50, 257
83. A.W. Sheldon, "ORTEPA Workshop, Toxicology and Analytics of the Tributyltins - The Present Status, Berlin, May 15-16, 1986", ORTEP Association, Vlissingen-Oost, NL, 1987, p. 101
84. K.M. Crofton, K.F. Dean, V.M. Boncek, M.B. Rosen, L.P. Sheets, N. Chernoff, L.W. Reiter, Toxicol. Appl. Pharmacol., 1989, 97, 113
85. M.D. Nemec, "A Teratology Study in Rabbits with TBTO, Final Report, Project No. WIL B0002", WIL Research Laboratories Inc., Ashland, OH, 1987 Property of Aceto Corp., Atochem N.A. and Schering AG
86. Bio/dynamics Inc., Report on project no. 88-3261, East Millstone, N.J., 2990 (Property of Aceto Corp., Atochem N.A. Inc. and Schering AG)
87. P.A. Baaijens, "ORTEPA Workshop, Toxicology and Analytics of the Tributyltins - The Present Status, Berlin, May 15-16, 1986", ORTEP Association, Vlissingen-Oost, NL, 1987, p. 191
88. C.R. Meyer, C.R. Buncher, R. Gioscia, J. Dees, "Oceans '87 Proceedings, International Organotin Symposium, Halifax, Nova Scotia, Canada, Sept. 28-Oct.1, 1987", The IEEE Service Center, Piscataway, NJ, and The Marine Technology Society, 1987, Vol.4, p. 1432
89. R. Balogh, "Personal communication", Schering AG, Bergkamen, 1987
90. S.J. Blunden, A.H. Chapman, Environm. Technol. Letters, 1982, 3, 267
91. R.J. Maguire, Appl. Organomet. Chem., 1987, 1, 475
92. R.J. Maguire, J.H. Carey, E.J. Hale, J. Agric. Food Chem., 1983, 31, 1060
93. A.E. Slesinger, I. Dressler, in M. Good (ed.): "The environmental chemistry of three organotin chemicals", Report of the Organotin Workshop, University of New Orleans, New Orleans, LA, 1978, p. 115
94. R.J. Maguire, R.J. Tkacz, J. Agric. Food Chem., 1985, 33, 947
95. D. Barug, Chemosphere, 1981, 10, 1145

96. R.J. Maguire, P.T.S. Wong, J.S. Rhamey, Can. J. Fish. Aquat. Sci., 1984, 41, 537
97. D. Barug, J.W. Vonk, Pest. Sci., 1980, 11, 77
98. T. Stein, K. Kuester, Z. Wasser Abwasser Forsch., 1982, 15, 178
99. K. Fent, Mar. Environm. Res., 1989, 28, 477
100. L.E. Hallas, J.J. Cooney, Dev. Ind. Microbiology, 1981, 22, 529
101. D. Liu, K. Thomson, Bull. Environ. Contam. Toxicol., 1988, 36, 60
102. M. Polster, K. Halacka, Ernährungsforschung, 1971, 16, 527
103. P.T.S. Wong, Y.K. Chau, O. Kramar, G.A. Bengert, Can J. Fish. Aquat. Sci., 1982, 39, 483
104. E.A.M. Mathijssen-Spiekmann et al., "Onderzoek naar de toxiciteit van TBTO voor een aantal zoetwaterorganismen", Report 668118001, Rijksinst. voor Volksgezondheit en Milieuhygiene, Bilthoven, 1989
105. ABC-Laboratories, "Acute toxicity of bis(tributyltin) oxide to Daphnia magna", Report 38308, Columbia MO, 1990 (Property of Atochem NA and Schering AG)
106. ABC-Laboratories, "Chronic toxicity of bis(tributyltin) oxide to Daphnia magna", Report 38310, Columbia MO, 1990 (Property of Atochem NA and Schering AG)
107. J.R. Dharia, B.M. Gavande, S.K. Gupta, Toxicol. Environm. Chemistry, 1989, 24, 149
108. ABC-Laboratories, "Acute 96-hour low-through toxicity of bis(tri-n-butyltin) oxide to rainbow trout (Oncorhynchus mykiss)", Report 38306, Columbia MO, 1990 (Property of Atochem NA and Schering AG)
109. R.C. Martin, D.G. Dixon, R. J. Maguire, P.V. Hodson, R.J. Tkacz, Aquatic Toxicology, 1989, 15, 37
110. W. Seinen, T. Helder, H. Vernig, A. Penninks, P. Leeuwangh , Sci. Total Environ., 1981, 19, 155
111. ABC-Laboratories, "Acute 96-hour flow-through toxicity of bis(tri-n-butyltin) oxide to bluegill (Lepomis macrochirus)", Report 38307, Columbia MO, 1990 (Property of Atochem NA and Schering AG)
112. R. Abel, "European policy and regulatory action for organotin based antifouling paints", Toxic Substance Division, Department of the Environment, London (Submitted for publication)

113. R.S. Carr, J.L. Hyland, T. Purcell, L.J. Larson, L.T. Brooke, "Oceans '87 Proceedings, International Organotin Symposium, Halifax, Nova Scotia, Canada, Sept. 28-Oct.1, 1987", The IEEE Service Center, Piscataway, NJ, and The Marine Technology Society, 1987, (Proceedings addendum) 6 ppüs
114. H. Plum, Internat. Environ. Safety, 1982, Dec., 4pp

Assessment of the Efficacy of Wood Preservatives

A.F. Bravery

TIMBER DIVISION, BUILDING RESEARCH ESTABLISHMENT, WATFORD WD2 7JR, UK

1 SUMMARY

Within the UK the use of wood preservatives is regulated under the Control of Pesticides Regulations, 1986. These Regulations require that only approved pesticides shall be sold, supplied, used, stored or advertised. The approvals process is administered by the Health and Safety Executive (HSE), Pesticides Registration Unit and includes assessment of efficacy as well as of safety in use.

The HSE assesses efficacy of wood preservatives on the basis of evidence submitted in support of specific claims by the manufacturers. It is for the applicant to set out the target pests against which he claims his product is efficacious and to provide the test or other evidence in support of those claims. The present UK procedures are flexible and in principle admit any objective evidence, although most evidence is likely to derive from British or European Standard method of test.

Under new procedures currently being developed in Europe in support of the Construction Products Directive, the philosophy and mechanism for assessment of wood preservatives will have important differences. New European Standards are well advanced which set clear performance requirements for each of 5 specific end-use situations defined as Hazard Classes. Only existing and newly developed European Standard Methods of test are permitted for establishing the efficacy profile of a wood preservative, and the target organisms relevant to each hazard class are clearly defined.

2 INTRODUCTION

Within the UK, wood preservatives are regulated under the Control of Pesticides Regulations 1986 (CPR), enacted under the Food and Environment Protection Act 1985 (FEPA). These regulations have superseded the former voluntary agreement known as the Pesticides Safety Precautions Scheme (PSPS). Under the PSPS specific industries including those producing wood preservatives entered into a Voluntary Agreement with Government Departments not to produce and market in the UK, certain pesticidal products without prior registration and approval of safety in use.

Under the Control of Pesticides Regulations 1986, in effect the voluntary agreement became mandatory and it is now an offence to sell, supply, use, store or advertise certain specified types of pesticides including wood preservatives, without prior approval. The Regulations allow for general requirements on sale, supply, storage, advertisement and use to be set out in 'Consents'.

The regulations have been framed primarily to ensure the safety of the environment and of persons who may come into contact with the specified pesticidal products. However under the Regulations products must be efficacious and assessment of their efficacy is an integral part of the approvals process.

Wood preservatives are normally applied to extend the service life of a given commodity to several decades. It therefore becomes necessary to have available techniques and procedures which can allow prediction of performance well into the future, for it is clearly impracticable to await the results of service or field trials before such products can be assessed for suitability.

3 REQUIREMENTS OF THE REGULATIONS

The Food and Environmental Protection Act 1985 describes a pesticide as any substance, preparation or organism prepared or used, among other uses, to protect plants or wood or other plant products from harmful organisms, to regulate the growth of plants, to give protection against harmful creatures, or to render such creatures harmless. The precise definition, together with details of pesticide products which fall outside the scope of the legislation, is given in Regulation 3 of the Control of Pesticides Regulations 1986.

Wood preservatives are clearly and specifically

cited as covered under the scope of the Regulations and therefore must have approval before they can be sold, supplied, used, stored or advertised.

There are 3 categories of approvals:

1) an experimental permit to enable testing and development to be carried out to establish the potential of a product and to provide data needed for a higher level of approvals,

2) a provisional approval which may be granted for a stipulated period of time, during which any data required for full approval can be obtained,

3) a full approval granted without restriction on its period of validity.

Products granted an experimental approval cannot be advertised or sold, whereas those granted provisional or full approval can be sold, supplied, used, stored and advertised. Products which have provisional or full approval are listed in the Reference Book 500 'Pesticides'[1], published annually and new approvals, changes and revocations are published regularly in 'The Pesticides Register'[2]. Responsibility for administration of the approvals procedure for wood preservatives rests with the Pesticides Registration Section of the Health and Safety Executive (HSE) at Bootle. Applications are received by HSE and pass through a process of internal scrutiny to a Scientific Sub-Committee for submission to the Advisory Committee on Pesticides which provides the advice to the Ministers under whose authority Approvals are granted.

The process of assessment and approval is based on consideration of the claims of applicants in respect of specific, formulated products, although evidence submitted in support of claims may take into account behaviour and properties of individual active ingredients.

4 PRINCIPLES FOR ASSESSMENT OF EFFICACY

Although the approvals process comprises assessment both of safety in use and of efficacy, only efficacy will be considered here.

As indicated previously, the basis for efficacy assessment in the UK is the claim of the applicants. The starting point for the application process therefore is the setting out of minimal descriptive information concerning the name and nature of the product, its specific intended label claim and its intended field of application (see Figure 1).

The effectiveness of a wood preservative in use depends upon several different factors, individually and in combination, in addition to the basic biological activity of the biocides themselves together with other ingredients of the formulation. In particular it depends upon the method of application of the preservative and the service environment for the treated wood, both of which are provided for in the application procedure (Figure 1). The permanence of the preservative product in the wood during service is also very important and this must be assessed in the laboratory, field tests or from evidence derived from service experience.

Wood preservatives are expected and intended to confer long-term protection measured in decades. Whilst service records and field trials are valuable and might be preferred it would be impracticable to wait on results from service experience before assessment can take place. Therefore methodologies are required which can enable early prediction of likely efficacy in service. In practice, data from laboratory tests will usually be required to provide the main core of objective evidence, especially for new products or variations of products not already well established in service.

5 METHODS OF TEST

Laboratory methods of test need to be rapid, reliable, reproducible and relevant to the intended field of use. Whilst many laboratory methods of test for evaluating activity against specific biological targets already exist as British and European Standards, there are also many non-Standard methods[3]. There are also Standard methods for determining the effects on efficacy of loss of preservative from treated wood due to evaporation ageing and to leaching in water. The main range of methods already available as European Standards is described in BS.6559 : 1985[4] but additional methods are now in place as a result of continuing activity in the European Standards Technical Committee CEN/TC38 'Durability of wood and wood products'. The nature, significance and ecology of the main target pests have already been described elsewhere in this book.

Figure 1 **Application for efficacy assessment of a wood preservative under the FEPA Regulations**

1	APPLICANT
2	ADDRESS

3 PRODUCT DETAILS (AS SOLD)

NAME

LABEL CLAIMS

Tick boxes as appropriate

CONCENTRATE		READY FOR USE	

SOLID	
LIQUID	
PASTE	
OTHER (BRIEF DESCRIPTION)	

4 FIELD OF APPLICATION *Tick one box only*

TIMBER PRE-TREATMENT	
IN SITU/REMEDIAL	
PROTECTION OF GREEN TIMBER BEFORE CONVERSION OR USE	

Figure 1 continued

5	FULL DETAILS OF FORMULATION (AS SOLD)

6	METHOD OF PREPARATION FOR USE (IF CONCENTRATE)

7 METHOD OF APPLICATION — Tick boxes as appropriate

BRUSHING AND SPRAYING	
LOCALISED IMPLANT/INJECTION	
IMMERSION	
DIFFUSION	
DOUBLE VACUUM/LOW PRESSURE	
HIGH PRESSURE	
OTHER (BRIEF DESCRIPTION)	

8 CLAIMED EFFICACY AGAINST — Tick boxes as appropriate

BLUE STAIN FUNGI (IN SERVICE)	
MOULDS AND SAPSTAIN FUNGI (GREEN TIMBER)	
WOOD-ROTTING FUNGI	
MARINE BORERS	
COMMON FURNITURE BEETLE	
HOUSE LONGHORN BEETLE	
DEATH WATCH BEETLE	
POWDER POST BEETLE	
TERMITES	
OTHER INSECTS (SPECIFY)	

Figure 1 continued

9 SERVICE ENVIRONMENT OF TREATED TIMBER — Tick boxes as appropriate

IN WATER	
IN THE GROUND	
ABOVE GROUND BUT NOT COVERED OR COATED	
ABOVE GROUND AND COVERED	
ABOVE GROUND AND COATED	
WHEN FRESHLY FELLED, IN FOREST OR SAWMILL	

Table 1 here summarises the main categories of pests for which tests of efficacy are available and which need to be considered for particular end-use situations. It also identifies where new methods of test are needed or are under development. Tables 2 to 4 list the methods of test available in relation to pretreated timber for the particular service environments and for remedial applications in service, in the form presently adopted for UK Efficacy Assessment.

BS.6559 : 1985[4] describes the important philosophy behind the laboratory methods of test which have been developed and adopted as European Standard methods for deriving indicators of efficacy in service. All the methods have certain key features in common. These include -

1) wood species - a wood substrate is employed to allow for the important interactions which may affect efficacy and to ensure that the organism is in the correct physiological condition. The species of timber are selected to be inherently highly susceptible to the target organism so as to maximise the challenge for the preservative under test. The wood species must be easily treated to give uniform distribution of the product under test and must be suited to the relevant target organism (e.g. a hardwood for hardwood selective organisms).

2) method of treatment - uniform distribution of the test preservative is achieved by employing vacuum impregnation of wood samples, except where the absorption characteristics of a surface application is specifically required. The application procedures are not intended to represent accurately any particular commercial process but they are precisely defined for reproducibility and reliability.

3) choice of organism - fungal and insect test species are selected to be economically important, common in practice (or representative of those that are), easy to handle in the laboratory, aggressive and consistent in their behaviour and response. The types of organisms that attack wood have been described in an earlier Chapter. Those which are the subject of Standard methods of test are listed in Table 1 of this Chapter.

Table 1 **Schedule of European Standard tests for assessing effectiveness of the wood preservative formulations**

Test requirement	EN Standard number Preventive efficacy Laboratory	Field	Eradicant efficacy
Insects			
Woodworm (Anobium)	EN21	na	EN48
	EN49	na	pr EN
Longhorn (Hylotrupes)	EN46	na	EN22
	EN47		
Powder Post (Lyctus)	EN20	na	EN273
Termites (Reticulitermes)	EN117	(EN252)	na
	EN118		
Fungi			
Decay (Basidomycetes)	EN113	EN252(G)	nst
	pr EN	pr EN330	
Blue stain in service	EN152	nst	na
Soft rot fungi	pr ENV	EN252(G)	na
Marine organisms			
Mixed (natural)	na	EN275	na
Physio-chemical conditioning			
Evaporative ageing	EN73	na	na
Leaching	EN84	na	na

Key: na = not applicable; nst = no standard test; (G) = ground contact; (OG) = out of ground contact

Service environment of treated timber	Equivalent hazard category	Example	Moulds	Stains	Wood-rotting fungi	Marine borers	Insects
Above ground and covered (dry)	1	Domestic roof timbers	NA	NA	NA	NA	EN insect ± EN 73
Above ground and covered (risk of wetting)	2	Roof timbers subject to condensation	BS 3900: part G6	EN 152: part 1 or 2	EN 113 ± EN 73	NA	EN insect ± EN 73
Above ground, not covered but coated	3	Painted external joinery	NA	EN 152: part 1 or 2	EN 113 ± EN 73	NA	NA
Above ground, not covered or coated		Fence rails	BS 3900: part G6	EN 152: part 1 or 2	EN 113 ± EN 73 and EN 113 + EN 84	NA	NA
In the ground, ground contact, or in fresh water	4	Fence posts, transmission poles	NA	EN 152: part 2	EN soil bed EN 252	NA	NA
In sea water	M	Marine piling	NA	NA	EN 275	EN 275	NA
Freshly felled, in forest or sawmill	—	Green timber	NST	NST	NST	NA	NST

Table 2 Tests available for the assessment of efficacy of pretreatment wood preservative systems

Insecticidal systems

Method of application	Target insect	Eradicant action	Preventive action
Spray, brush or mayonnaise	Anobium	EN48/EN370	EN21 or EN49 ± EN73
	Hylotrupes	EN22	EN46 or EN47 ± EN73
	Xestobium	NST	NST
	Lyctus	EN273	EN20 ± EN73
	Termites	NST	EN117 or EN118 ± EN73
Localised injection	Xestobium	NST	NST
Gas/vapour diffusion	All wood borers	NST	NA
Smoke	All wood borers	NST	NA

Table 3 Tests available for the assessment of efficacy of remedial/in situ wood preservative systems

Fungicidal systems

Method of application	Eradicant action	Preventive action
Spray or brush		
– internal constructional timber		
anti-mould	NST	BS 3900:Part G6
anti-rot	NA	NST
– external timbers		
anti-mould	NST	BS 3900:Part G6
anti-stain	NST	EN152 Part 1
anti-rot	NA	NST
Mayonnaise		
– internal constructional timber	NST	NST
Injection or implants		
– external joinery	NST	NST
Bandages		
– poles and heavy structural timber	NST	NST
Gas/vapour diffusion		
- poles and heavy structural timber	NST	NST

Table 4 Tests available for the assessment of efficacy of remedial/in situ wood preservative systems

4) pre-conditioning - tests at present only address changes in the treated wood which might occur due to the two most significant physico-chemical factors influencing longevity of effect in service; loss of the preservative by evaporation and water leaching. The techniques prescribed are attempts to represent in the laboratory what happens in practice by standardising the main experimental variables of temperature, time and air speed, to provide a unified basis for a common and consistent challenge. The field is a developing one and improvements are under investigation. It is already known for example that thermal effects and free-radical induced chemical changes can reduce preservative efficacy but no standardised methods of test yet exist to assess these factors.

The biological methods of tests are written to be appropriate to determination of the efficacy of wood preservatives applied either as pre-treatments to prevent biological attack, or as remedial treatments to disinfest or to eradicate existing attack. At present eradicant tests apply only to wood-boring beetle species because remedial treatments are commonly applied in practice to eliminate existing infestations of these insects. No Standard eradicant fungal tests exist because in general the strategy of attempting to eradicate fungi from wood in practice by applying fungicides is considered suspect and liable to serious abuse. It can be a valid practice in certain very specific situations and for these circumstances specialised techniques will be required in order to demonstrate efficacy.

In assessing a wood preservative formulation, the choice of target organism(s) will depend upon the perceived risks related to the intended conditions of use of the timber it is to protect in service. These perceived risks are the starting point for any assessment of the efficacy of a wood preservative as well as for the choice of the tests to be conducted. A rationale for classifying these risks was originally evolved by the European Homologation Committee. This has been further developed for the UK efficacy assessment scheme[5] and by the European Standardisation Committee for the durability of wood and wood products (Technical Committee CEN TC38). The rationale is based on describing a series of service environments for treated timber, called 'hazard classes'. Originally the UK favoured 6 such classes (Table 2) but has now accepted the CEN preference for 5.

Particular organisms predominate under the conditions described by the respective hazard classes. Since the hazard classes are essentially linked to predicable prevailing moisture conditions, they describe principally the increasing degree of risk from the different types of wood attacking fungi. This leads to an assessment system where tests against certain fungi would become obligatory, because of a defined high risk of attack, whilst other tests, including those against insects, remain optional because of a recognisably lower level of risk. Under such a scheme it is for the producer and/or specifier to decide and make known his claims or requirements. Broad spectrum activity of individual preservatives, that is activity against as wide a range of the organisms as possible, was once a generally accepted attribute and objective but is no longer sought in the same degree. Moves towards environmentally more benign chemicals have encouraged acceptance of a narrower spectrum of activity as an attribute in itself rather than a weakness. Decisions as to the principal target organism thus dictate the basic activity required. Broader spectrum activity, where desired, is increasingly achieved by mixing chemicals with specific activities, though this can bring difficult and expensive consequences in testing for efficacy assessment, as well as for toxicological evaluation.

Laboratory tests are designed and intended to establish the basic, innate toxic efficacy of the wood preservative against specific organisms under carefully controlled and reproducible conditions. Field tests are intended to enable the combined effects and interactions of the natural factors which influence efficacy to be tested - factors such as the weather (rain, sun, temperature) but also including the natural, mixed populations of biological agencies. Because these tests involve essentially long term exposure to practical conditions, they are regarded as service tests mainly for fully formulated commercial products. Some shortening of the test period is achieved by prescribing sample sizes which increase vulnerability to attack and by presenting the samples in a way which optimises attack, for example, by maximising the influence of wetting and the invasion opportunity for the attacking organisms.

The assessment of the preventive efficacy of wood preservative formulations has to be made from values derived from the relevant biological test. These values are either the actual quantitative amounts of the product established in the test as causing the appropriate level of mortality of the target organism, or they represent the threshold limits so-called, the 'toxic values'. The toxic values are two concentrations in the series used in the test, the one which just permitted continued attack and that next higher concentration which just prevented it.

The preventive efficacy tests in general can be used with the product formulated for commercial use, with developmental products or indeed to evaluate new biocidal chemicals in specific solvents or model formulations.

7 EVALUATION AND INTERPRETATION

The background and intentions of the UK Scheme for efficacy assessment has been described by Morgan et al[5]. Fundamental to the process is the requirement for the applicant to present a reasoned case and not just a compilation of miscellaneous data. Whilst evidence assembled for specific products has to be considered on a confidential basis, the principles for assessment are intended to be as open and rational as the available knowledge and experience will allow. Unfortunately very few of the biological tests have been devised or can be operated as "pass/fail" methods. Furthermore there are very few accepted performance targets for wood preservatives and no clearly established correlations between data from laboratory tests and those from field tests, service trials or practical experience. Setting rational criteria is further complicated by the fact that many of the preservatives that have been in widespread use, give differing performances in laboratory tests. This makes it almost impossible to provide a fully objective basis for assessment without recourse to arbitrary criteria which may be unfair or grossly permissive.

Some approaches to the development of an objective rationale were described by Morgan et al[5] and these rely heavily on the principle of internal reference preservatives based on products or active ingredients with a sufficiently long history of satisfactory performance in service. The procedures are still relatively complicated though derived from an admittedly simplistic logic. There is a continuing need for the introduction of a moderating procedure if there is not to be a tendency towards either fail-safe over specification of wood-preservatives or for potentially inadequate under-specification.

8 THE EUROPEAN APPROACH

The European approach to assessment of the effectiveness of wood preservatives currently being evolved by CEN Technical Committee TC38, only addresses wood preservatives for pre-treatment processes as required in the context of the Construction Products Directive. It does however go further than possibly any individual country in the world has gone before, in attempting to set out clear and unequivocal prescriptions for the testing and performance Standards.

The drafts issued for public comment through 1990 and 1991 have as their main objective the establishment of performance Standards which will facilitate free trade across the open European market intended to take effect from 31 December 1992. Compliance with the new performance Standards will be signified by the adoption of the European CE mark accompanied by descriptive labelling.

There has been a deliberate effort within the UK to ensure strong similarities between the interim UK approach and that being developed for Europe for the latter may well ultimately supercede the UK scheme. The CEN Scheme is complex and still being evolved so it cannot be set out in full here. The basic principles and structure have much in common with the UK Scheme already illustrated here in Table 2. There is a 5-tier hazard class system to designate the intended service environment and within each class the essential biological challenges are identified by a prescription of the minimum test requirements for each. The maximum amount of the preservative product which may be applied in the test is defined, together with the so-called 'biological reference value' which is the minimum amount of product deemed effective in the test. Additional testing options are provided together with their relevant application and performance criteria, and these tests are expected to be called-up according to the claims of manufacturers and/or requirements of specifiers.

The CEN suite of tests for efficacy assessment of a given wood preservative product will provide a series of 'biological reference values' for the relevant biological agencies in a given hazard class. The minimum amount of product for use in this service situation is deemed to be the 'critical value' which is the highest of the biological reference values thus obtained.

The CEN performance Standards for wood preservative products do not stand alone. They are part of a mutually

Table 5 European Standard methods of test for wood preservatives (at 1 March 1991)

European Standard No.	Title	British Standard No.
20-1	Wood preservatives-Determination of the protective effectiveness against Lyctus brunneus (Stephens) - Part 1 : Application by Surface treatment (Laboratory method)	BS.5217 : 1975
20-2	Wood preservatives-Determination of the Protective effectiveness against Lyctus brunneus (Stephens) - Part 2: Application by impregnation (Laboratory method)	-
21	Wood preservatives-Determination of the toxic values against Anobium punctatum (De Geer) by lrvoal transfer (Laboratory method)	BS 5218 : 1989
22	Wood preservatives-Determination of eradicative action against Hylotrupes bajulus (Linnaeus) larvae (Laboratory method)	BS 5219 : 1975
46	Wood preservatives-Determination of the preventive action against recently hatched larvae of hylotrupes abjulus (Linnaeus) (Laboratory method)	BS. 5434 : 1989
47	Wood preservatives-Determination of the toxic values against Hylotrupes bajulus (Linnaeus) Larvae (Laboratory method)	BS.5435 : 1989
48	Wood preservatives-Determination of the eradicant action against Larvae of Anobium punctatum (De Geer) (Laboratory method)	BS.5436 : 1989
49-1	Wood preservatives-Determination of the protective effectiveness against Anobium punctatum (De Geer) by egg-laying and larval survivals. Part 1 : Application by surface treatment	BS.5437 : 1977
49-2	Wood preservatives-Determination of the protective effectiveness against Anobium punctatum (De Geer) by egg-laying and larval survival (Laboratory method). Part 2 : Application by impregnation	
73	Wood preservatives-Accelerated ageing tests of treated wood prior to biological testing-Evaporative ageing procedure	BS.5761:Pt1.1989
84	Wood preservatives. Accelerated ageing of treated wood prior to biological testing. Leaching procedure.	BS.5761:Pt.1989

Table 5 continued

113	Wood preservatives-Determination of toxic values of wood preservatives against wood destroying Basidiomycetes cultured on an agar medium	BS.6009 : 1982
117	Wood preservatives-Determination of toxic values against Reticulitermes santonensis de Feytaud (Laboratory method)	BS.6239 : 1990
118	Wood preservatives-Determination of preventive action against Reticulitermes santonensis de Feytaud (Laboratory method)	BS.6240 : 1990
152-1	Test methods for wood preservatives-Laboratory method for determining the preventive effectivenss of a preservative treatment against blue stain in service - Part 1 : Brushing procedure	BS.7066:Pt1:1990
152-2	Test methods for wood preservatives-Laboratory method for determining the protective effectiveness of a preservative treatment against blue stain in service ... Part 2 : Application by methods other than brushing	BS.7066:Pt2:1990
212	Wood preservatives. Guide to sampling and preparation of wood preservatives and treated timber for analysis	BS.5666:Pt1:1987
252	Field test method for determining the relative protective effectiveness of a wood preservative in ground contact	BS.7282 : 1990

reinforcing set of standards which cover 1) the classification of hazard classes, 2) the classification and requirements for the natural durability of wood and wood products, 3) the minimum performance standards for wood preservatives and 4) the specification of preservative treated wood. In addition there is the large number of already published Standard methods of biological test, together with a series of new methods still under development. The list of published Standard methods is given in Table 5. Those published as draft or provisional Standards and those still under development are listed together in Table 6.

9 CONCLUSIONS

Assessment of the efficacy of wood preservatives presents some formidable difficulties not encountered in the assessment of most other pesticidal products. These difficulties arise for many reasons; 1) there is such a large range of damaging organisms exploiting the wide diversity of service conditions under which wood is used, 2) wood itself is a complex, highly orientated mixture of natural polymers which, through the different wood species has great variability in form, composition and properties, 3) wood preservatives have to be formulated in a variety of different ways to accommodate to the consequences of the above variables as well as to take account of their intended function either as a preventive or remedial product. For many uses of wood preservatives there are still no agreed Standardised methods of test. Interpretation of the data from laboratory, field and service tests is complicated because even well established products provide data which are difficult to use as reference Standards, as they do not have a constant relationship between preservatives concentration in use and its performance in the tests.

Rigid use of test data brings with it a risk of over- or under-specification of wood preservatives for service conditions. Public concerns for the environment and for the health and safety of individuals will not countenance over-specification and fitness for purpose legislation in its different forms, will not accept under-specification. A simple prescriptive assessment of the efficacy of wood preservatives has still to be fully realised.

Table 6 Draft European Standard methods of test for wood preservatives and new methods under development (at 1 March 1991)

Provisional No.	Title
prEN 273	Wood preservatives. Determination of the curative action against Lyctus Brunnens (Stephens) (Laboratory method)
prEN 275	Test method for determining the protective effectiveness of a preservative against marine borers
prEN 330	Wood preservatives-Field test for determining the relative protective effectiveness of a wood preservative for use under a coating and exposed out of ground contact L-joint method
prEN 335-1	Definition of hazard classes of biological attack - Part 1 : General
prEN 335-2	Definition of hazard classes of biological attack - Part 2 : Solid Wood
prEN 335-3	Definition of hazard classes of biological attack - Part 3 : Wood-based panels
prEN 350-1	Natural durability of wood - Part 1: Principles of testing and classification of the natural durability of wood
prEN 350-2	Natural durability of wood - Part 2: Natural durability and treatability of selected woodd species of importance in Europe
prEN 351-1	Preservatives-treated solid wood - Part 1 : Requirements for preservative-treated wood according to hazard classes
prEN 351-2	Preservative-treated solid wood - Part 2 : Sampling and analysis of preservative-treated wood
prEN 351-3	Preservatives-treated solid wood - Part 3 : Identification of preservative-treated wood
prEN 460	Natural durability of wood - Durability classes and hazard classes
-	Performance of wood preservatives as determined by biological tests - Part 1. Specification according to hazard classes
-	Performance of wood preservatives as determined by biological tests - Part 2. Classification and labelling
-	Test for determining the toxic efficacy against soft rotting microfungi and soil inhabiting micro-organisms
-	Method of test for determining the relative effectiveness of wood preservatives applied by surface processes

Table 6 continued

-	Wood-based panels : Resistance to fungal attack
-	Accelerated ageing by simplified leaching
-	Performance of biocides in timber - Part 1. Fixation method in water
-	Performance of biocides in timber - Part 2. Fixation method in air

REFERENCES

1. MAFF/HSE, Pesticides - 1990 Reference Book 500, HMSO, London 1990.
ISBN 0-11-2428:9-3.
2. The Pesticides Register. HMSO, London. ISBN 0-11-728659-1.
3. Bravery, A F, Methods for assessing the efficacy of fungicides for wood and wood products in 'Industrial Microbiological Testing'.
SAB Technical Series 23. Blackwell Scientific Publications, (1987).
ISBN 0-632-01793-7.
4. British Standards Institution, General Introductory document on
European (or CEN) methods of test for wood preservatives.
BS 6559:1985 HD/1/160 (1985).
5. Morgan, J W W, Assessing the efficacy of wood preservatives.
Rec Ann Conv Brit Wood Preserv Assoc, Cambridge, 1988.

Progress Towards European Standards in Wood Preservatives

F.W. Brooks

FOSROC LIMITED, TIMBER TREATMENTS DIVISION, FIELDHOUSE LANE, MARLOW, BUCKINGHAMSHIRE SL7 1LS, UK

1 INTRODUCTION

In a previous presentation on this subject, during the British Wood Preserving Association Annual Convention in July 1989, I reported that the first deadline set by the European Commission for the completion of European Standards on wood preservation as part of the preparation for the Single European Market (SEM) had just passed without agreement having been reached by the various working groups. I concluded that lecture by anticipating that one further year to obtain agreement was reasonable; that was by July 1990.

This presentation briefly describes the background of the SEM and gives the up-to-date position in the search for our particular ambition, which will permit wood preservatives and treated wood to be traded freely throughout all EC member states without needing expensive, time consuming, individual approvals to be given by each separate country. By way of a topical example of what the SEM will achieve I have had many conversations recently with UK timber joinery producers who have been appointed suppliers to the large Disneyland construction project in Paris. The timber treatment specifications for the joinery in this project refer to particular French Standards for preservative and method of timber treatment. We found that there were no facilities in the UK which could meet the letter of these specifications. Further research provided the name of a product available in France which did comply with the French Standards. Unfortunately this product was not registered in the UK by the Health and Safety Executive under the Control of Pesticides Regulations and so could not lawfully be imported and used in the UK. The SEM will cut through these obstacles so that specifications written in one member state must be based on European Standards which can be met in all member states.

2 SINGLE EUROPEAN MARKET

The creation of a single trading market in Europe was actually embodied in the original 1957 Treaty of Rome. Between 1957 and 1987 progress towards harmonisation was slow because each piece of legislation was written on the basis of setting objectives and writing detailed technical specifications. Also total agreement was required between member states before legislation could be enacted. This was difficult enough when only six countries comprised the European Economic Community. When the UK, Ireland and Denmark joined in 1973, followed by Greece in 1981 and then Spain and Portugal in 1986 the problems of reaching unanimous agreement became pragmatically insurmountable.

All these difficulties were swept aside by the Single European Act (SEA), which was the vision of Jacques Delors and drafted by Lord Cockfield, and which came into force on 1.7.1987. This Act is the agreement under which member states undertake to enact the necessary legislation to create a single trading market by 31.12.1992. The main provisions of the SEA, in order to speed up the process of achieving a single market, are the removal of the two earlier requirements, firstly, for complete harmonisation of technical specifications and secondly, unanimity in coming to decisions. The aim, now, under the new Act, is to obtain sufficient harmonisation, based on mutual recognition of each country's systems, and to adopt legislation based on majority voting principles. The SEA provided a timetable with a deadline of midnight on 31.12.1992 for the incorporation of 300 directives (European laws), which have subsequently been reduced to 279, which will complete the single trading market.

3 CONSTRUCTION PRODUCTS DIRECTIVE

The piece of legislation which affects our industry is the Construction Products Directive (CPD).[1] This was one of the first of the 'new approach' directives under the SEA and was signed on 21.12.1988. Implementation by member states is required by 27.6.1991 and this should be achieved in UK. The directive defines the essential requirements of constructions and states that a construction product is fit to trade, and may be traded throughout the Community without hindrance from technical and regulatory requirements, and without the need for re-testing or re-certification to meet national or local conditions, if it enables the construction to satisfy the relevant essential requirement(s) for a reasonable period. The essential requirements of constructions are defined as:

- mechanical resistance and stability
- safety in case of fire

- hygiene, health and environment
- safety in use
- protection against noise
- energy economy and heat retention

The directive applies to products which enable the essential requirements to be met. Products are deemed to satisfy the essential requirements if they conform to approved technical specifications. These will be mainly Harmonised European Standards (ENs) but may also be a European Technical Approval (ETA) or a Non-Harmonised Technical Specification recognised at Community level. In the situation of the wood preservation industry it was decided in 1988 by a mandate from the European Commission that Harmonised European Standards will be technical specifications adopted by the European Standards Organisation, CEN, and it is at this level that intense activity is being made by industry, Government departments, etc, to achieve a majority view on how these standards should be written.

4 CEN

During 1987 the Technical Board of CEN set up a Programming Committee to advise on the need and format for European Standards in the construction industry, arising from which a number of CEN Technical Committees (TCs) were either formed, or had their original mandates extended, to prepare all necessary standards to enable timber to be used in compliance with the CPD. These are:

CEN/TC 124 Structural timber.
CEN/TC 112 Wood based panels.
CEN/TC 103 Timber adhesives.
CEN/TC 38 Durability of wood and wood based panels.

TC 38 has in fact been in existence since the early 1960's. Its original mandate was to prepare methods for the biological testing of wood preservatives.

Subsequently it was recognised that a further TC, designated CEN/TC 175, was needed to produce EN's for sawn timber and saw logs for non-structural purposes, e.g., fencing, joinery. These new TC's began work in the early part of 1988 and each created separate working groups to produce the specific standards which were identified by the TC's.

It is not in the remit of this paper to amplify the work of TC 124, TC 112, TC 103 or TC 175. The work of TC 38 is of greatest importance to the wood preservation industry.

5 CEN/TC 38

The first meeting of CEN/TC 38 with its expanded remit was in April 1988 and the main decisions of the meeting were the definition of the basic philosophy of how wood preservatives and treated wood should be specified and then the creation of the necessary working groups to produce the ENs.

Working Group 1 Hazard classes

It was agreed that the way to specify durability of timber was to identify timber in service according to the biological hazards to which it is exposed. Traditionally in the UK we have considered timber durability by reference to the commodity. Within a territory such as the UK this does not give rise to any difficulties of interpretation. On the geographical scale of Europe there are significant climatic differences which give different hazards to particular commodities, e.g. timber in ground contact in Mediterranean climates is subject to termite attack not found in northern Europe. Also window joinery in climates in the north west Europe will remain wet longer than joinery in the dry interior. Therefore standards based on commodities would not necessarily be appropriate in all cases.

WG1 was created to define the minimum number of classes to which timber is exposed according to the severity and type of biological hazard. Table 1 shows the present draft classification for solid timber which will be published as EN 335, 'Wood and wood based products - definition of hazard classes of biological attack'.

Timber in hazard class 1 will always have a moisture content below 18 per cent and therefore will not be attacked by wood destroying fungi. It will be liable to attack by insects and this will depend on the timber species and geographical location. Four insect species need to be considered: woodworm (*Anobium punctatum*), house longhorn beetle (*Hylotrupes bajulus*), Lyctus powder post beetle (*Lyctus brunneus*) and termite species. The specifier will decide which insects are present, and whether they represent a significant risk to the construction, and will choose an appropriate preservative and treatment, or alternatively a timber species of adequate natural durability.

The remaining hazard classes 2-5 represent situations where the moisture content of the timber will achieve 20 per cent for varying periods of time and so the timber will be at risk from fungal attack. Hazard classes 2 and 3 cover situations where timber is not in ground contact and a judgement needs to be made about the length of time the component remains wet. Hazard class

Table 1 Definition of biological hazard classes for solid timber

Hazard class	Situation in service	Description of exposure	Wood moisture content
1	Above ground, covered (dry)	Permanently dry	Permanently < 18%
2	Above ground, covered (risk of wetting)	Occasionally wet	Occasionally > 20%
3	Above ground, not covered	Frequently wet	Frequently > 20%
4	In contact with ground or fresh water	Permanently wet	Permanently > 20%
5	In salt water	Permanently in salt water	Permanently > 20%

4 is for timber in ground contact or in fresh water, and hazard class 5 includes timber in marine conditions.

A separate classification will be made of hazard classes for board materials in conjunction with CEN/TC 112 and this will be issued as another part of EN335.

Working Group 2 Natural durability. The responsibility of WG2 is to prepare EN350, 'Wood and wood based products - natural durability of wood'. This will provide the timber specifier with information on the natural durability of wood against fungi (both in and out-of-ground contact), wood destroying insect larvae, termites and marine organisms. In addition it is intended that a classification of timber species according to permeability (treatability) will be prepared. This information will enable the specifier to choose a timber species of sufficient natural durability to give adequate performance when exposed to the different biological agencies in each of the hazard classes.

Working Group 3 Performance of treated timber. This group is concerned with specification of treated timber which ultimately will form EN 351, 'Durability of wood and wood based products - preservative treated wood', and will replace the existing British Standards BS5268 : Part 5 'Code of practice for the structural use of timber. Preservative treatments for constructional timber', and BS5589 'Code of practice for preservation of timber'. The outcome of this group will therefore be most relevant to those concerned with industrial commercial timber treatment in the UK.

It is important to note that remedial treatment of timber and treatment to protect against sap stain are not considered to come within the scope of the Construction Products Directive and so are not a part of the deliberations of the CEN committees.

This WG is making slow progress and it is proving difficult to find a level of agreement between the many ways which exist in different European Countries to specify treated wood.

The present draft of EN 351 specifies treated timber according to a series of penetration requirements which are summarised as follows:

P 1	None
P 2	None
P 3	3mm lateral sapwood 40mm axial sapwood
P 4	6mm lateral sapwood
P 5	6mm lateral sapwood 50mm axial sapwood
P 6	12mm lateral sapwood
P 7	Full sapwood
P 8	Full sapwood 6mm exposed heartwood
P 9	20mm lateral or full sapwood (lesser)
P 10	20mm lateral of full sapwood (greater)

These cover a range from no requirement for timber treated by a superficial method to full sapwood or 20mm heartwood penetration. Detailed guidance is not given about which penetration level should be applied to particular timber commodities. This is the responsibility of the specifier taking into account the climate, desired service life, function in the building e.g. whether load bearing or not, standards of maintenance, etc. Associated with the penetration requirement is a loading of preservative in the defined penetration zone. This loading is a characteristic of the preservative itself and not simply the active ingredient(s).

Working Group 4 Performance of wood preservatives. WG4 is responsible for defining the performance required of wood preservatives under laboratory and/or field test conditions in order for them to be approved for the treatment of timber in the hazard classes defined by WG1. The biological tests defined are the European Standards which have been published by CEN arising from the original work of CEN/TC 38. In carrying out its work WG4 has highlighted areas where performance needs to be specified but where no EN exists. In consequence further WG's (numbers 5-10) have been created by CEN/TC 38 to produce the necessary tests.

The tests which will characterise a wood preservative will be selected according to the hazard class and type of organism to which the timber is exposed in service, and according to the method of application of the wood preservative, from the following categories of biological tests:

- insect resistance after treatment either by surface application or impregnation.
- basidiomycete fungal resistance after treatment by surface application or impregnation.
- field tests for performance in ground contact, out-of-ground contact and in the marine environment.
- artificial ageing (evaporation and water leaching).

EN tests are also available for determining the resistance of treated timber to blue stain organisms in service and these could be called by the specifier as additional tests if local conditions so require. The list of existing EN standard tests available, or being developed, to characterise a preservative is as follows:

Insects	Lyctus	EN 20, 273
	Hylotrupes	EN 22, 46, 47
	Anobium	EN 21, 48, 49, 370
	Termite	EN 117, 118
Fungi	Basidiomycete	EN 113
	Soft rot	EN ...
Field tests	"L" joint	EN 330
	Stake test	EN 252
Marine	Marine borer test	EN 275
Ageing	Evaporation	EN 73
	Leaching	EN 84
Others	Blue stain in service	EN 152
	Sampling for analysis	EN 212
	Fungal resistance, board materials	EN ...

In addition to identifying the schedule of tests required for preservatives used to treat timber in each hazard class the working group will also define the minimum performance levels for a wood preservative, from each EN test, to demonstrate suitability for each hazard class. This will be the basis of required retention in the treated timber standard EN 351.

6 COMPARISON OF THE NEW EUROPEAN SCHEME WITH EXISTING BRITISH STANDARDS

A complete picture of the new structure, when all working groups have finished their work is summarised in the following diagram.

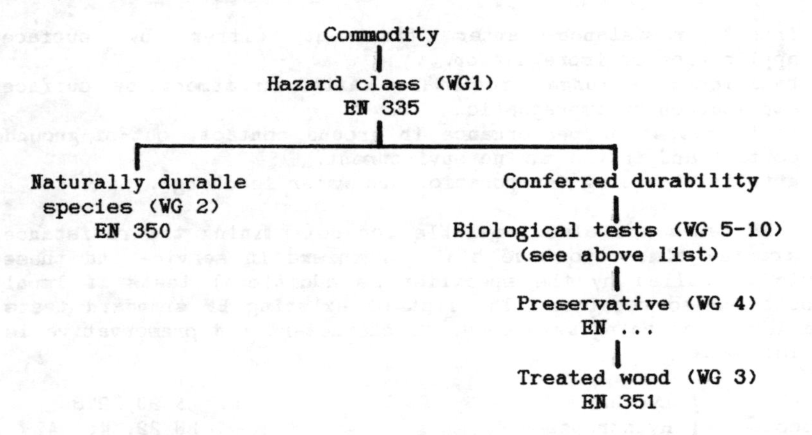

This shows that a timber commodity must be assigned to a particular hazard class, given in EN 335. The specifier then has a choice of either using a suitably naturally durable timber (EN 350) or using chemical treatment with a preservative according to EN... (not yet assigned) to produce treated timber according to EN 351 based on a penetration and retention requirement.

This structure is completely different from the system which is currently in place in UK, which has developed over the last twenty years, in which a wood preservative is specified according to its active ingredients and is applied to the timber by a specified process to suit the end requirements. The UK British Standards structure, illustrated by the more important standards, is as follows:

Preservative (active ingredients)	+	Process	→	Commodity
				BS 5589
BS 114: Part 1		BS 144: Part 2		
				BS 5268: Part 5
BS 4072: Part 1		BS 4072: Part 2		
				BS 1990
BS 5707: Part 1		BS 5707: Part 3		
				BS 1186

The wood preservation industry in the UK will have to change its approach and adopt the European philosophy when the EN's are published. Each nation has an obligation to withdraw all conflicting national standards when new EN's are adopted. Therefore these familiar BS numbers will be but affectionate memories before long.

7 ADOPTION OF THE CPD AND USE OF CE MARK

I have not given any detailed dates for publication of the EN standards to support the CPD. The original deadlines proposed by the CEN working groups were unrealistic and depended on having no disagreements between member states on matters of philosophy or detail. Present expectations are that the standards being produced by WG1 to 4 will be finalised this year and adopted by member states as official ENs during 1992. In the meantime the CPD will be implemented into national law in June 1991 and in the interim period between then and the adoption of the EN standards existing national standards will be deemed to satisfy the essential requirements of the CPD, but only in the country in which the national standard applies. In other words until the EN's are agreed there cannot be a free market for construction products in Europe.

Once all standards and laws are in place then producers of construction products which comply with all aspects will be able to use the CE mark of conformity which is the passport for product acceptance in the European market. The directive does not make the mark mandatory. However, products which do carry the mark are presumed to be fit for their intended use and they may not be rejected on technical grounds by any member state. This will gladden the heart of all UK producers of wood preservatives and treated timber who wish to export, but are prevented from so doing at the moment by the mass of legislative restrictions which apply in other European Countries. The corollary of this of course is that the UK market will also be open to companies from other European countries. As far as countries from other continents are concerned, which export to UK based on British Standards, then they too will have to ensure that products comply with the new European Standards by the time that the conflicting BS's are withdrawn.

8 HEALTH, HYGIENE AND THE ENVIRONMENT

It is necessary to return to the detail of the Construction Products Directive and consider again the essential requirements.

The use of the CE mark to support a construction product means compliance with all relevant essential requirements. The discussion so far in this paper has only concerned the first essential requirement, mechanical resistance and stability, and this is the only requirement for which CEN is mandated to produce standards. It will, in addition, be necessary for the manufacturer to ensure that wood preservatives and treated timber are not damaging to hygiene, health and the environment, which is the third essential requirement of constructions conforming to the CPD.

This requirement means that the construction must be built in such a way that it will not be a threat to the hygiene or health of the occupants or neighbours, in particular as a result of the giving-off of toxic gas or the presence of dangerous particles or gases in the air.

The details of compliance with this essential requirement have not yet been elucidated. A draft interpretative document has been produced by the Standing Committee on Construction.[2] This explains in greater detail the meaning of the essential requirement, but it does not state how products will or will not comply, and these instructions still have to evolve, probably from committees in the European Commission. It is important that the industry, through European organisations such as the European Wood Preservative Manufacturers' Group and the Western European Institute for Wood Preservation keep abreast of these deliberations as they could have as much impact on products as the new EN standards for durability.

CEN/TC 38 is keeping contact with this subject through the most recently created working group WG11. This group is entitled, 'Permanence of active ingredients in treated timber', and is working on methods of test for measuring losses of preservative into the atmosphere or water. Such information may be needed to demonstrate compliance with essential requirement III. CEN/TC 38 also has a correspondence group on environment and toxicology which can pass on the views of CEN to other decision making bodies within the EC.

9 ATTESTATION OF CONFORMITY.

The final subject I wish to raise, and again only very briefly, as much discussion still has to take place at the level of the Commission before decisions are made, is the Brussels-speak phrase of attestation of conformity. In other words what does the product manufacturer have to do to show that his product complies with its technical specification. The CPD makes provision for four levels of attestation of conformity from full independent third party certification to manufacturer's declaration of conformity. The four levels are:-

1. Certification by approved body.
2. Manufacturer's declaration of conformity + certification of production by approved body.
3. Manufacturer's declaration + initial type testing by approved body + manufacturer's production control.
4. Manufacturer's declaration + initial type testing by manufacturer + manufacturer's production control.

The choice of scheme will be related to the importance of the product in enabling the essential requirements to be met and the production control needed, and will be decided by the Standing Committee but based on four criteria formulated in 4 questions:

Q1. is the part played by the product with respect to at least one of the essential requirements important, medium or little?

Q2. are small variations in the product's characteristics or properties likely to endanger significantly the serviceability and the working life of the work?

Q3. are these characteristics or properties likely to vary significantly as a result of small variations in manufacturing processes and in manufacturing parameters?

Q4. are such variations in manufacturing parameters difficult to control?

These four questions can be posed on the following decision tree.

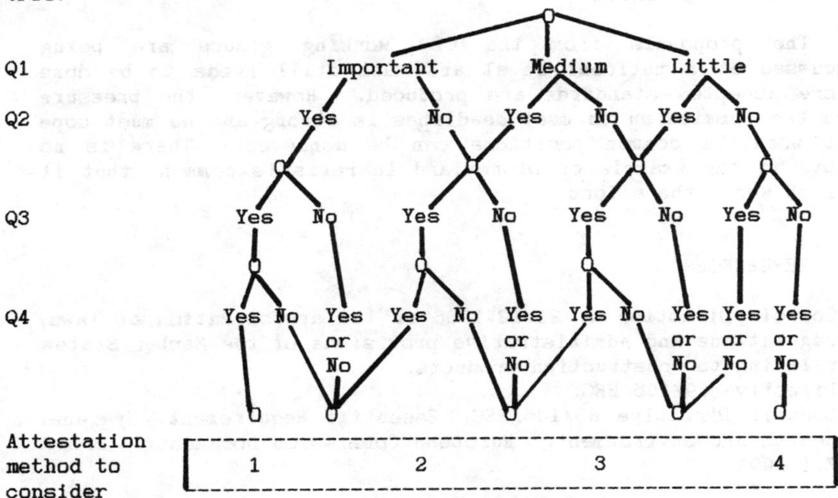

If we postulate that the answers to the questions are as follows:

Q1. influence of the product on essential requirements is 'important'.
Q2. small variations in product are 'not significant'.
Q3. small variations in production process are 'not significant'.
Q4. variations in manufacturing are 'not difficult to control'.

then the choice of attestation of conformity to be considered this between 2 and 3, that is a manufacturer's declaration of conformity and either third party factory production control or manufacturer's production control in conjunction with initial type testing by a third party. These decisions will be made by the Standing Committee during 1991 or 1992

10 CONCLUDING REMARKS

This completes my 'snapshot' of the present activity of the wood preservation industry's contribution to a United Europe which is accurate as of March 1991. It must be emphasised that this is still an interim statement of the development of European Standards and products complying with the Construction Products Directive. No doubt there will be changes in the future and the record of this Conference will not describe the requirements as they ultimately unfold.

The proposals from the CEN working groups are being discussed at a national level and much still needs to be done before accepted standards are produced. However, the pressure from the Commission to meet deadlines is strong and we must hope that workable common positions can be achieved. There is no doubt, if the example of Disneyland in Paris is common, that it will be worth the effort.

REFERENCES

1. Council Directive of 21.12.1988 on the approximation of laws, regulations and administrative provisions of the Member States relating to construction products.
Directive 89/106/EEC.
2. Council Directive 89/106/EEC Essential Requirement, Hygiene, Health and Environment. European Commission Document TC 3/017 7.1.1991.

Preservatives for the Market — or the Market for Preservatives?

C.R. Coggins

RENTOKIL LTD., FELCOURT, EAST GRINSTEAD, WEST SUSSEX RH19 2JY, UK

INTRODUCTION

The need for effective wood preservatives has always existed. Accounts exist of methods employed, for example, by the ancient Greeks, the Romans, the Chinese and Alexander the Great in ancient Persia. These efforts were almost inevitably in vain because the ancients did not understand the causes of wood decay and their use of sea water and various oils was doomed to failure.

The industrial revolution brought demands for the effective preservation of poles, sleepers and mine timbers, providing its own answers in the form of creosote oil as a by-product of the processing of coal tar and the beginnings of the metal salt preservatives in the form of zinc or mercuric chloride. The remarkable extent of early research and development of wood preservatives is reflected in a treatise on wood preservation of timber by W.Chapman in which he said:

> "Almost every chemical compound of any plausibility has been suggested and submitted but the multiplicity of opinions forms nearly an inextricable labyrinth"
>
> Treatise on the Preservation of Timber,
> W Chapman, 1817.

Simple soaking and hot-and-cold open tank methods were used alongside the vacuum-pressure method patented by Bréant in 1831 and utilised by Bethell in his famous 1838 patent covering the impregnation of timber with creosote oil by the full-cell process.

Also in 1838 Boucherie developed his process for the treatment with copper sulphate of freshly-felled trees for use as poles using a simple but effective sap-displacement process.

Upon these foundations was built the wood preservation industry we know today.

THE EMPIRICAL AGE

The shortcomings of the main metal salt preservatives - zinc and mercuric chloride and copper sulphate were recognized early this century.

Zinc chloride has fungicidal properties but it is easily leached from timber. Furthermore, it decomposes in water releasing hydrochloric acid which destroys the wood and any metal fittings. Treated sleepers were found to have a life of less than five years.

Mercuric chloride-treated timber suffered from failures in ground contact and its highly poisonous nature and other drawbacks led to its demise.

Copper sulphate cannot be used in steel plants because of its corrosive nature and failures of treated wood in contact with chalky soils also led to its demise. Nevertheless its use on a huge scale reflected the need for even partly effective preservatives - between 1860 and 1910 around 5,500,000 poles were treated with copper sulphate by the Boucherie process.

Of course right up to the present day creosote has continued in extremely effective use, refined and produced in accordance with strict specifications. Its oily nature and odour, however, tends to limit its use to heavy duty industrial components such as sleepers and poles. This has helped to maintain the pressure to develop cheap and effective water-borne metal salt preservatives for many end-uses.

One of those who responded to that need was a young German engineer, K.H.Wolman who had gone to Silesia in 1900 where major collieries were being exploited. His attention was drawn to the problem of preservation of pit timber and he set his mind to the development of impregnating salts or compounds which would not harm wood or iron and which would resist leaching from the wood. The history of the company which Wolman founded and which still bears his name is in the main the history of wood preservation at least up to the period from 1900 to the second world war.

A sequence of patents granted to Wolman shows how he exploited first the fungicidal properties of fluorides and nitrated phenols. He then identified that 2% - 5% of chromates and dichromates could advantageously be added to these substances to prevent corrosion of iron. In 1921 the need for preservation of timber in termite-infested areas led to the addition of arsenic and in the 1930's perhaps the most significant development - the increase of the chromium content to 35% to achieve fixation of the water-soluble salts in the treated wood.

Meanwhile outside Germany in the 1930's similar important developments were taking place. Gilbert Gunn patented several variants on the copper/chromium theme which he called Celcure. Kamesan in 1933 patented a new preservative in which he added copper sulphate to a previously patented mixture of arsenic and chromium compounds - the first CCA which he called 'Ascu'. The standard set by this and, particularly, the later CCA formulations, of performance, cost-effectiveness, ease of use and lack of collateral problems show what a serendipitous development Kamesan made. 40 years of subsequent intensive research have failed to produce another waterborne preservative with benefits approaching those of CCA.

THE MODERN ERA

In the period from 1950 to the present day the benefits of the copper/chromium/arsenic mixture coupled with its refinement in the form of the mixtures enshrined in British Standard 4072, Boliden

K33 from Sweden (AWPA Type B) and ultimately the Type C version in the USA have led to its domination of the world water-borne wood preservative markets.

Other types have had local importance greater than CCA, notably the copper/chromium/boron and copper/chromium/fluoride preservatives in Germany,Switzerland and Austria, but they are part of the same generic group as CCA's, have common roots and essentially do the same job.

In the same period there has grown up a market need for preservatives which can be used to treat machined joinery sections and other precision wooden building components without affecting dimensions or surface characteristics. This has been satisfied with the light organic solvent preservatives based on really quite simple mixtures of highly effective organic biocides - DDT,dieldrin,lindane,synthetic pyrethroids, PCP, zinc soaps,tributyltin compounds - in solvents of the white spirit type.

The period is well characterised by statistics for production of preservative treated wood in Sweden since 1950. Fig.1 shows the growth of treatments by water-borne preservatives at the expense of creosote and the birth of the organic solvent business.

THE MARKET NOW AND THE FUTURE

The 1990's finds the manufacturer of wood preservatives facing major difficulties associated with the growth of awareness of the requirements for health, safety and environmental protection.

Legislators are introducing new controls over the marketing and use of chemicals of all types but pesticides in particular. A debate on the background to and relevance of current developments to the wood preservation industry is out of place here but wood preservatives are subject now in some countries to the same controls as those imposed on agricultural pesticides spread directly onto the land and water in great quantity.

One particular problem is the 'catching-up' exercise being carried out by product approval authorities on 'old' products like CCA and its variants, creosote and some of the earlier organic solvent preservatives. In this exercise the lists of requirements for characterisation of animal toxicity and eco-toxicity of new products are compared with data available on existing products. Data-gap call-ins are then implemented which may require the expenditure of millions of pounds to fill. Such matters lead to the question can the old preservatives survive?

The present market for wood preservatives is characterised by:

- an existing population of treatment plants using a very limited range of highly cost-effective preservatives in accordance with well established treatment parameters and handling procedures

- a timber trade on which our industry lives which suffers low profit margins and often views preservation as cost-added rather than value-added

- strong environmental pressures form both regulatory authorities and ginger groups against all of the major preservative groups

The opportunities for research and development groups to bring new preservatives to this market are therefore tremendous today, perhaps mirroring the opportunity facing the young K.H.Wolman in Silesia in 1900. What is not open, however, is the cheap route to developments which he enjoyed. The full gamut of toxicity testing is required before products may be introduced to the market. New active ingredient toxicity testing costs run into millions of pounds against which the costs of registering a formulation based on a new active substance (£30,000 in the UK) seem to pale into insignificance, yet these are greatly in excess of anything seen by our industry before.

New performance standards for wood preservatives are about to be introduced in Europe. These are product-specific and mean that the 'type' specifications like BS4072 will no longer have any role. The old or existing products may not have been tested to the new requirements and may therefore need re-testing to the same regime as entirely new preservatives.

Treatment companies will have to bring their plants up to standards of health, safety and pollution control which are being both changed and raised by regulators. This is necessary, of course, but it raises the cost of staying in or entering this business to a very high level. The prospects for the sale of new preservatives with more desirable properties are limited by the higher prices demanded by increased raw material costs and high testing costs.

Thus what is needed from research and development is a combination of support for the existing products to ensure the continued availability to society of these cost-effective products where they can be used without compromising health and safety or the environment and the bringing forward of new formulations and techniques which provide realistic options for the timber trade, perhaps most importantly offering the opportunity to extend the uses of timber or timber-based materials with the added value of preservation.

The old marketing adage that the company which has the right product at the right place at the right time and at the right price will be a successful company will undoubtedly prove true in the 1990's and beyond. In this context the right product will not always be the new product.

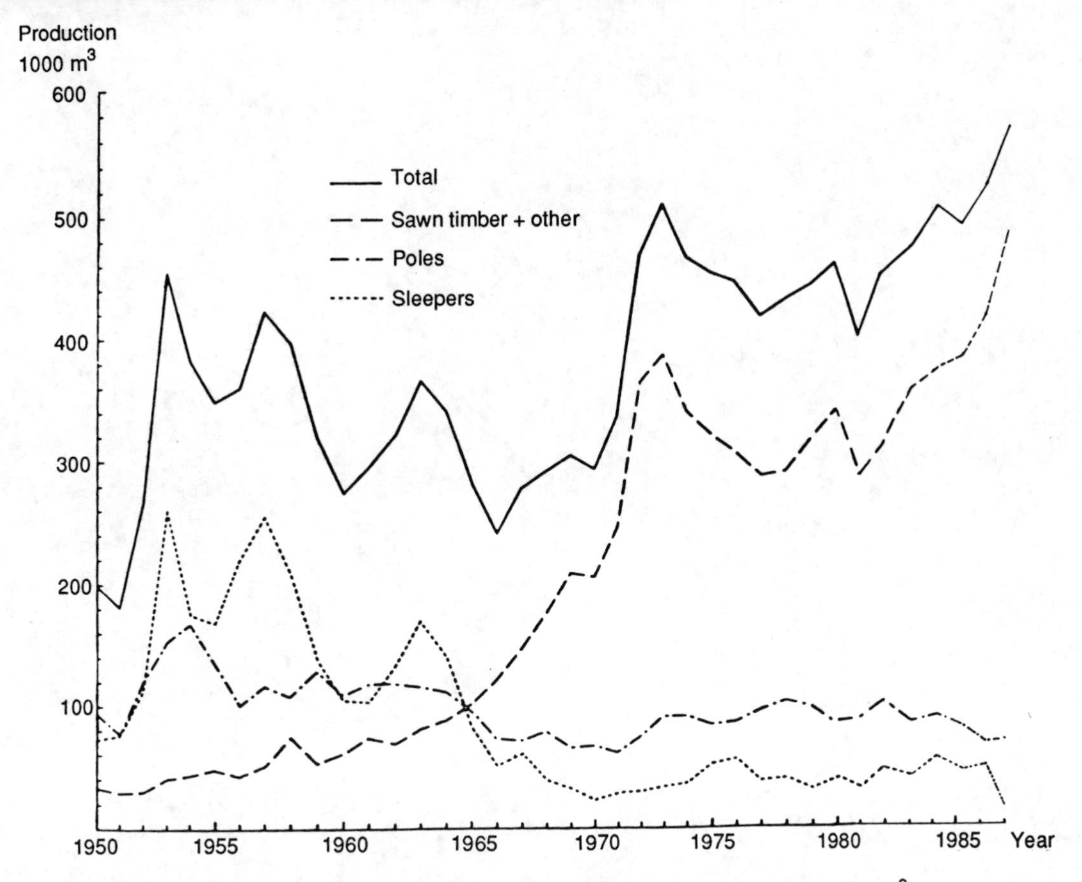

Figure 1 Production of preservative-treated wood in Sweden 1950-1987 by commodity, m^3.

Vapour Phase Treatment of Wood with Trimethyl Borate

P. Vinden, R.J. Burton, and A.J. Bergervoet

WOOD TECHNOLOGY DIVISION, FOREST RESEARCH INSTITUTE, ROTORUA, NEW ZEALAND

1 ABSTRACT

Changes in the wood processing industry, including increasing raw material costs, fluctuations in market demand, higher interest rates and stock holding costs, are demanding more automation and a move to "on-demand" processing rather than production of stock for anticipated demand. A new generation of wood treatment technology is described involving vapour phase application of wood preservatives. The treatment of solid wood has been integrated into the kiln drying process, whereas the treatment of panel products is carried out at ambient temperature and at the board equilibrium moisture content.

2 INTRODUCTION

New Zealand has the highest per capita consumption of preservative-treated timber in the world, because our building regulations require virtually all timber in dwellings to be preservative treated. The treatment industry began to take off in New Zealand in the mid-1950s, after a Committee of Inquiry set up by Government to consider the efficacy of borax/boric acid treatment. The Committee accepted boron as a wood preservative, but also established the Timber Preservation Authority because of concern at the lack of co-ordination between Government agencies and private sector groups involved with wood preservation.

Production statistics for boron-treated timber illustrate rapid growth until the mid-sixties. Boron preservatives successfully captured the low decay and insect hazard market, e.g., timber framing. They have since resisted competition from copper-chrome-arsenic (CCA) and light organic solvent preservatives (LOSP).

The method of boron treatment – momentary immersion and block diffusion storage for up to 8 weeks[1] – has remained virtually unchanged since the mid-1950s. Traditionally, boron preservative is applied to rough-sawn timber, mainly because of the difficulty in achieving adequate uptake of preservative and anti-sapstain chemical in gauged material. Subsequent gauging after diffusion storage results in a 15% loss of timber and an even higher percentage loss of the boron salt originally applied to the timber surface. Untreated shavings have value as a feed stock for composite products, whereas boron-treated shavings can be a liability.

Patented improvements to this treatment[2] have been introduced into the industrial process using a thickened boron treatment, "Diffusol"®. This formulation provides higher uptake of preservative, which facilitates the treatment of gauged timber rather than rough-sawn timber, the use of cold application rather than a heated bath, and the use of lower solution concentrations. Within-charge retention variability is also lower which, together with gauging before treatment, facilitates a 25% saving in time to achieve specification requirements. Over a period of 18 months, 50 plants have adopted thickened boron treatment in New Zealand.

Diffusion treatment with boron has a number of disadvantages. The diffusion storage period requires vast stocks of timber to be held – in some cases up to one-third of a mill's total annual production of boron-treated timber. This presents problems in a fluctuating market, and hardly lends itself to "just-in-time" production.

A number of techniques have been developed to accelerate boron treatment, involving steam conditioning and pressure impregnation[3,4], high-temperature diffusion[4,5], kiln drying, and pressure impregnation using the "Lite process".[6] All these processes could meet specification requirements but could not compete with the simplicity and low capital cost of the diffusion process.

Preliminary work into vapour treatment[4,7] indicated that it was possible to achieve effective boron treatment after kiln drying of timber. This fitted well with an international trend towards using dry framing. The cost of vapour boron treatment is competitive with the traditional dip/diffusion process, followed by drying. Patent applications were lodged for VBT treatment and a major joint research programme into vapour phase treatment of timber and panel products was initiated between the Forest Research Institute, New Zealand, and the Department of Biology, Imperial College, London. This paper summarises some of the results of that programme of work.

3 FUNDAMENTAL ASPECTS OF VBT

The chemistry involved in trimethylborate (TMB) vapour phase treatment of wood is illustrated in Equation 1.

$$B(OCH_3)_3 + 3H_2O \rightarrow H_3BO_3 + 3CH_3OH \quad (1)$$

The hydrolysis of TMB to boric acid and methanol is nearly instantaneous, with 1.7 kg TMB reacting with 0.9 kg water producing some 1.0 kg boric acid. Although the reaction is well known, we verified the relationship with moisture in wood, using wood wool. The potential boric acid loading is strongly related to moisture content of wood (Figure 1).

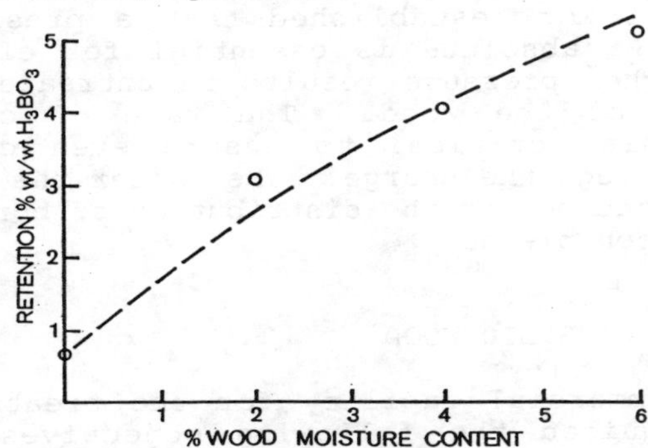

Figure 1 Effect of wood moisture content on boric acid retention

The sensitivity of TMB to moisture provided the focus for early work. A high wood moisture content results in the deposition of boric acid on the surfaces of timber rather than penetrating the wood. Laboratory trials identified five main factors influencing depth of chemical penetration and treatment variability.

- Wood substrate permeability
- Wood moisture content at time of treatment
- Treatment vessel pressure
- Effectiveness of vapour generation and distribution
- Treatment temperature

There are differences in the penetration pathways between panel products and solid wood. In panel products the gas penetrates through the matrix rather than fibres. This facilitates treatment at ambient temperatures and equilibrium moisture content.[9] The pathways in solid wood are restricted to ray tissue and tracheid/tracheid interconnections.

Reduction of internal pressure in the treatment vessel serves three purposes. Although the moisture content of the wood has been reduced to approximately 6%, evacuation of the vessel/kiln serves to remove further moisture which would otherwise react prematurely with TMB. The boiling point of TMB is approximately 60°C whereas the Azeotrope (TMB + Methanol) is 55°C. A reduction in pressure obviously reduces the boiling point and aids vapourisation of the chemical. TMB is flammable. Removal of oxygen also minimises risk of fire.

Experimental work established that a pressure of less than 15 kPa absolute is essential for effective treatment. Higher pressure results in untreated areas in the centre of the wood. The rate of chemical injection is also critical to ensure even chemical distribution through the charge. The faster the rate of vapourisation, the better the distribution of boric acid within and between pieces.

4 TREATMENT OF SOLID WOOD

Design of a commercial facility for the treatment of solid wood required the following objectives to be fulfilled.

- Environmental attractiveness
- Safety in use
- Financial attractiveness
- Compatibility where possible with current practices in the industry
- Simplicity
- Short processing times
- Total processing time compatible with shift operations

A schematic layout of the process is illustrated in Figure 2.

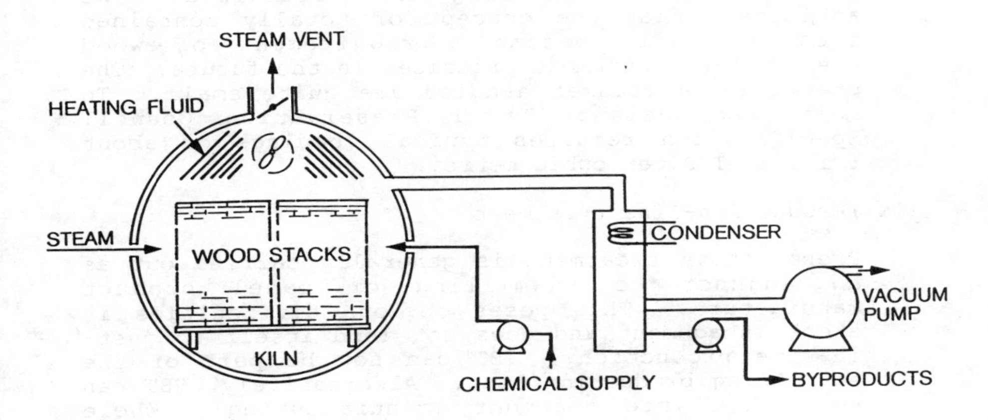

Figure 2 Schematic layout of Vapour Boron process

Features of the plant include:

(i) Total integration of drying, preservative treatment, and steam conditioning

High temperature kiln drying (HTD) of radiata pine uses a conventional schedule of 120°C dry bulb and 70°C wet bulb.[10] Drying takes about 22 to 24 hours. The kiln is then sealed and evacuated to below 15 kPa (absolute) and the chemical injected. A short period of time is allowed for the solution to equilibrate. The plant is then evacuated through a condenser to remove reaction products – mainly water and methanol. Total treatment time is about 1 hour. The final stage is steam conditioning. This is a normal HTD operation to relieve stresses in the wood after drying and takes about 4 hours for 50-mm-thick material. Steam conditioning also hydrolyses any residual TMB to boric acid so that no vapour escapes into the atmosphere after removal from the cylinder.

(ii) Totally self-contained treatment

Treatment operations ranging from chemical delivery to removal of by-products for regeneration into more TMB can be contained in intelligent containers with no operator contact with the chemical. There is no chemical mixing. Treatment is effected by injecting the required volume of chemical needed to achieve the desired preservative loading. Spillage and excess chemical handling are eliminated. We anticipate that the concept of totally contained treatment will become a requirement of wood preservation treatment practices in the future. The preservative volumes handled are quite small. To meet New Zealand Timber Preservation Council Specification requires typical loadings of about 6 litres TMB per cubic metre.

(iii) Production-line treatment

Preservative treatment is generally carried out as an adjunct to sawmilling or panel product manufacture. This poses obvious difficulties in stock management and does not lend itself to just-in-time production. VBT can now be part of the sawmilling production line. Alternatively, VBT can be linked into secondary manufacturing. Whole sections of houses or composite beams could be treated just prior to despatch.

(iv) Totally automated processing

Combining the drying and preservation operation reduces manual involvement. There is less stock handling and lower risks of stock damage with the closer co-ordination of stock processing. Automated control systems like DRYSPEC[11] which provide "intelligent" kiln operation can be modified to control the treatment operation, thus eliminating operator error. The consistency achieved between operations provides built-in quality control.

(v) Production timed to output needs

Timber is dried and treated ready for shipping to the customer in under 2 days rather than the traditional 8–10 weeks required for boron diffusion. The resulting reduction in storage requirements and in working capital tied up in stock presents obvious commercial and practical advantages, as does elimination of the need to guess market demand in 3 months' time. Three levels of chemical loading can be applied to provide insect, fungicidal, or fire resistance.

(vi) Low cost

Process cost is competitive with that of any other process producing dry-boron-treated wood, as illustrated by the internal rate of return (Table 1). VBT treatment is substantially cheaper than processes involving pressure impregnation because a second drying is eliminated.

5 COMMERCIAL TRIALS

The plant used for commercial trials had a capacity of approximately 150 000 litres and was used for high-temperature drying of charges of approximately 50 m^3. Research was conducted in two stages:

(i) Optimisation of kiln drying schedules
(ii) VBT treatment

Results of the drying trials demonstrated that drying to low moisture contents on a large scale was achievable and with good through-kiln uniformity. Wood degrade at 4 to 6% moisture content was not significant.

Table 1 Summary of unit costs (US$) and internal rates of return*

Plant size (m^3/day)	25	80
Vapour Boron		
Total start-up cost ($)	379,800	858,230
Drying/treatment cost ($/$m^3$)†	36.96	35.04
Diffusion/Drying		
Total start-up cost ($)	417,708	970,163
Drying/treating cost ($/$m^3$)†	32.74	30.25
Internal rate of return (%)		
Vapour Boron	20.5	38.1
Diffusion/drying	18.2	29.6

* The calculations assume a green timber cost of US$154/$m^3$ and selling price of $203/$m^3$. The costs include land, forklift movement, and energy.

† A time-cost element is not included here.

The results of preservative treatment are summarised in Table 2 and are better than can be achieved by any other currently available preservative treatment process in terms of uniformity within the kiln and even distribution through the cross-section of timber.

A series of commercial trials illustrated the importance of reducing the internal pressure prior to vapour generation. Data obtained from the commercial trials also fitted the models generated in laboratory trials.

The insertion of treatment between drying and steam conditioning worked extremely well. Both the gas generation technique and the recovery operation performed satisfactorily.

Table 2 Preservative retention analysis of 100 x 50-mm radiata pine sapwood

	Position in kiln			
	Front	Front/ Middle	Middle/ Back	Back
Mean cross-section retention (% (H_3BO_3) w/w)	0.35	0.32	0.33	0.34
minimum	0.34	0.26	0.30	0.29
maximum	0.37	0.39	0.36	
Mean core retention (% (H_3BO_3) w/w)*	0.36	0.42	0.46	0.40
minimum	0.28	0.31	0.40	0.37
maximum	0.42	0.50	0.51	0.46
Core : cross-section retention ratio	1.02	1.31	1.40	1.20

* Middle ninth of cross-section.

The higher core loadings reflect slightly higher wood-moisture contents in the cores due to moisture gradients.

6 CONCLUSION

Detailed laboratory research and full-scale commercial trials have demonstrated the viability of VBT treatment of solid wood. An audit of the commercial operation and an economic analysis confirm the environmental and financial benefits of the development. Treatment of solid wood was the most difficult operation to undertake. Research is now in hand on panel products. Full preservative treatment of boards can be achieved at ambient temperature and equilibrium moisture content. Research effort will now concentrate on the treatment of refractory or difficult-to-treat wood species and on mechanisms for fixing boron in the wood.

7 ACKNOWLEDGEMENTS

The authors would like to acknowledge assistance from K. Nasheri and D.R. Page in undertaking the work described in this paper.

8 REFERENCES

1. K.M. Harrow, N. Z. Timb. J. For. Rev., 1955, 2(3): 28-39; (4): 38-41.

2. P. Vinden, J.A. Drysdale and M. Spence, Thickened boron treatment, International Research Group on Wood Preservation, 1990, Doc. No. IRG/WP/3632.

3. A.J. McQuire, New Zealand Forest Service, New Zealand Forestry Research Notes No. 29, 1962.

4. P. Vinden, A. Fenton and K. Nasheri, Options for accelerated boron treatment: a practical review of alternatives, International Research Group on Wood Preservation, 1985, Doc. No. IRG/WP/3329.

5. A.J. McQuire and K.A. Goudie, N. Z. J. For. Sci., 1972, 2(2), 165.

6. P. Vinden, A.J. Bervergoet, K. Nasheri, P. Cobham, N.P. Maynard and M.A. Brown, 'A Timber Treatment Plant',. New Zealand Patent Application No. 234829, 1990.

7. R.J. Burton, A.J. Bergervoet, K. Nasheri, P. Vinden and D.R. Page, Vapour phase preservation of wood, International Research Group on Wood Preservation, 1990, Doc. No. IRG/WP/3631.

8. P. Vinden, R.J. Burton and T. Vaioleti, 'Gaseous or Vapour-phase Treatment of Wood with Boron Preservatives', New Zealand Patent No. 220, 816, 1988.

9. P. Turner, R.J. Murphy and D.J. Dickinson, Treatment of wood-based panel products with volatile borates, International Research Group on Wood Preservation, 1990, Doc. No. IRG/WP/3616.

10. D.H. Williams and J.A. Kininmonth, New Zealand Forest Service, FRI Bulletin No. 73, 1984.

11. W.R. Miller and S.G. Riley, Proceedings of the Wood Drying Symposium, Upgrading Wood Quality Through Drying Technology, F. Kayiham, J.A. Johnson and W.R. Smith (Eds), Seattle, 1989, 169.

The Performance of Alternative Antisapstain Compounds

G.R. Williams and D.A. Lewis

HICKSON TIMBER PRODUCTS LIMITED, CASTLEFORD, WEST YORKSHIRE WF10 2JT, UK

1 INTRODUCTION

The disfigurement of freshly felled logs and sawn timber, due to the rapid invasion by mould and sapstain fungi, can significantly reduce the yield and value of both hardwood and softwood timber species. This degrade results from the high moisture and nutrient status of freshly felled timber on which the fast growing mould and sapstain fungi are part of the primary colonizing microflora. Post harvest control of these organisms can be effected, quite simply by rapidly reducing the moisture content of the timber by, for example, kiln drying. Unfortunately, in many instances there can be delays between conversion and drying or, kiln drying may be economically non-viable, and a chemical treatment therefore becomes necessary in order to protect the timber during this susceptible period.

Antisapstain compounds have been defined as; Prophylactic chemicals[1] for the protection of green timber during the seasoning period.

For many years, the chlorinated phenols, particularly sodium pentachlorophenoxide, (NaPCP), have dominated the market for anti-sapstain chemicals by virtue of their cost-effectiveness and good performance. NaPCP is water soluble and applied to the timber at the final stage of processing by simple methods such as immersion, deluge or by pack dipping.

2 THE SEARCH FOR ALTERNATIVE COMPOUNDS

Increasing environmental pressure towards both the end user and producers of NaPCP has resulted in an extensive search for alternative products. A wide range of potential biocides have been examined including many existing agricultural fungicides, paint film biocides, algicides and bacteriocides. The characteristics of a suitable product include:

(i) a wide spectrum of activity

(ii) low toxicity

(iii) ease of handling and preparation and

(iv) cost effectiveness

Reference to the literature on this subject, however, emphasises the variability in performance encountered in many tests and a consideration of the various factors which govern the overall performance on antisapstain products is highlighted in this paper.[2,3,4]

3 THE PERFORMANCE OF ALTERNATIVE PRODUCTS

Williams has compiled a list[4] of some of the more common biocides tested for this application. Many of these will be familiar to users of biocides in other industries such as in water treatments, pharmaceuticals or agricultural pesticides. However, a number of these compounds have predominated in many studies, including the alkyl ammonium compounds,[5] the benzothiazoles particularly 2-thio cyanomethylthio benzothiazole (TCMTB),[6] thiocyanates such as methylene bis-thiocyanate (MBT),[4] iodine containing compounds including 3-iodo propanyl butyl carbamate (IPBC),[7] and organo-metallic compounds from which copper-8-quinolinolate (Cu-8) has shown potential.[2] Indeed all of the above compounds have been incorporated into commercial formulations and promoted as alternatives to NaPCP.

Factors Affecting Performance

Spectrum of Activity. The most obvious factor governing the performance of a biocide is not only its absolute activity but also its spectrum of activity towards a range of organisms. Early in the testing programme for alternative products, a number of active ingredients were found to provide adequate effectiveness towards staining fungi, but reduced effectiveness towards mould fungi. One organism which has caused particular concern has been the green mould *Trichoderma*. In some cases, a stimulation of mould growth has been observed,[8] over and above those levels recorded on untreated material; Williams suggested that this resulted from the suppression of less tolerant competitive organisms which thereafter allow the tolerant mould to dominate the population.[4] A number of products have been developed to overcome these problems which are based on mixtures of active ingredients with a broader spectrum of activity. Examples are products based on MBT and TCMTB such as Hickson ANTIBLU 3739 or IPBC and Quaternary ammonium compound mixtures.

In some instances, protection against early colonising decay fungi such as Phlebiopsis may also be required, particularly where timber is to be stored for longer periods of time. This is a difficult area however, and the protection of timber for excessive periods (greater than 6 months) under conditions of poor storage with little or no drying will require treatment using wood preservative compounds and not antisapstain compounds. This is not often recognised however and all too often unrealistic performance is expected from products which are not designed for this end-use.

The Formulation of Alternative Biocides. Formulation aspects can be particularly important to the performance of antisapstain compounds. This relates not only to the physical stability of a formulation but also its inherent activity towards target fungi at the wood surface. Many of the readily available organic biocides have low water solubilities and are, therefore, suitable only as water-emulsifiable or dispersible products. The stability of these solutions can be affected by a number of variables including water temperature, water hardness and solution contamination including the presence of high levels of sawdust in the treatment vessel.

It is important that these factors are thoroughly investigated during the development of alternative products, bearing in mind the potential variability which can be encountered in use at different locations. These factors can often be investigated in the laboratory although it is very difficult to simulate in-service conditions. Clearly there is no real alternative to extensive mill testing when investigating these parameters.

The efficacy of the active ingredients within a product can also be affected by formulation. Dickinson emphasises the importance of formulation to the performance of products based on copper-8-quinolinolate; the most recent formulations of this product contain the active ingredients in a solubilised form.[2] Similar results were obtained by Lewis et al who found that the formulation of MBT and TCMTB was important to the overall effectiveness of these compounds.[6] In particular, evidence suggests that alteration of both the emulsifier, solvent or co-solvent can directly affect both the stability and efficacy of the product.

Interaction of Biocide and Wood Substrate. The interaction of the components within a formulation and the wood substrate can significantly affect the performance of antisapstain compounds. Williams has described the difference in penetration properties between an MBT based product and NaPCP.[4] Treatment with NaPCP is characterised by a rapid precipitation to PCP in the acidic sap at the wood surface and, thereafter, limited penetration of the timber. In contrast, MBT formulations can be readily diffusible having significantly greater penetration properties. The treatment concentrations must take account of this effect or else surface depletion can occur resulting in product failure. ANTIBLU 3739 (Hickson Timber Products Limited) contains both MBT and TCMTB in the formulation, the latter component is of lower mobility and, therefore, provides surface protection.

Interaction of the biocide and the wood substrate can also affect the efficacy of the treatment solution. If during treatment the organic biocide is strongly adsorbed onto the wood substrate then the result can be a gradual depletion of active ingredient. This has been demonstrated for a number of active ingredients including alkylammonium compounds, TCMTB and some isothiazolones and is commonly known as solution stripping. This effect can be largely prevented by careful formulation using particular types and levels of emulsifier. Some degree of adsorption will be an advantage with antisapstain treatments by providing permanence and this is particularly important where the treated timber remains exposed to rainfall before it has dried.

Timber Species and Quality. The susceptibility of various timber species to degrade by mould and sapstain fungi is closely related to its natural durability. This applies both to hardwood and softwood timbers.

For example, Lewis et al found Spruce to be markedly less susceptible to degrade than either Scots pine or Corsican pine during field tests in the UK.[6] This of course, would not preclude the use of antisapstain chemicals to treat Spruce, but indicates that a lower level of biocide would be required to afford the same degree of protection. The converse is true for highly susceptible hardwood species such as rubberwood (Hevea brasiliensis). Indeed, such is the problem with rubberwood that log spraying with both fungicide and insecticide is recommended to prevent degrade. Thereafter, the most successful means of preventing further degrade is by pressure treatment with antisap-stain compounds to ensure adequate penetration and internal protection of this highly susceptible timber.[9]

A delay in conversion following felling of softwood species can also significantly reduce the effectiveness of an antisapstain treatment due to the increased microbial potential within the substrate. This is discussed more fully below.

Closely related to the above factors are the differences in local microflora. This has been highlighted in a number of areas. Rose indicated that the species of Trichoderma harzianum causing degrade on Pinus radiata in Chile was thought to be particularly tolerant to a wide range of organic biocides.[10]

Environmental Conditions and Post-Treatment Handling. The performance of any biocide is governed by the micro-environment in which it must be effective. Both the prevailing environmental conditions and handling procedures can significantly alter this micro-environment thereby changing the performance characteristics of a given compound. The effect of temperature is relatively straightforward in that below an optimum value, increasing temperature will result in increasing microbial development resulting in a greater overall hazard. This also applies to timber moisture content whereby a reduced rate of drying will invariably lead to reduced effectiveness of treatment and it is this factor which is most affected by handling procedures. In particular, the stacking format has been shown to affect the performance of antisapstain compounds. Lewis et al recorded a considerable increase in the susceptibility of untreated timber, and a corresponding reduction in performance of treated timber which was close stacked rather than open stacked.[6] Dickinson emphasises that in areas such as Portugal which have mild and damp winters then this constitutes a severe hazard if the timber is close stacked and the drying process therefore retarded.[2]

This effect can be exaggerated considerably where there is a long period of wet storage and such is the situation where large volumes of pallet timber are transported in the cargo hold of ships to distant markets. This constitutes a severe biological hazard since a ships hold often contain significant quantities of water and provides a warm and humid environment which is ideal for the development of a range of micro-organisms. A more effective treatment system is required for these situations and this is generally achieved simply by increasing the treatment solution concentration. However, this practice is often constrained by factors such as increased cost and dilute solution stability. The chlorinated phenols are well established 'wood preservative' compounds and are well suited to protecting timber in this hazardous environment. This results not only from their broad spectrum of activity but also their biological persistence. This persistence is one of the main reasons why they are considered to be environmentally unacceptable. Manufacturers of more modern organic biocides claim that their products are 'biodegradable' and as such the persistence of organic biocides in antisapstain treatments is an area which has not been fully researched and is contributory to the failure of treatments under these adverse conditions.

Recent work has demonstrated that the organic molecule TCMTB can be readily degraded by bacteria and yeasts isolated from freshly felled timber.[1] Results have indicated that some micro-organisms were able to detoxify up to 50% of the active ingredient in liquid culture after only 5 days incubation. Further work by Briscoe et al has also demonstrated the detoxification of a range of organic biocides used for antisapstain treatments under conditions which would be expected to prevail in timber.[11] Detoxification by bacteria isolated from fresh sapwood was recorded for TCMTB, MBT, Copper-8-quinolinolate and copper naphthenate.

More recently, work reported by Wallace et al has implicated the role of fungi such as Trichoderma in further biodetoxification of organic biocides. Results indicated that this organism is capable of detoxifying solutions of MBT at sub-toxic concentrations.[12] Similar results were reported by Briscoe et al using the sulphamide based compounds, Preventol A4S and A5.[11]

The failure of antisapstain compounds during prolonged wet storage may therefore involve initial detoxification by bacteria and yeasts, followed by the growth of tolerant fungi such as Trichoderma which may further degrade the biocide. This opens the way for further wood degrade by basidiomycete decay fungi if poor storage conditions are maintained for an extended period.

These studies highlight the importance of good storage conditions following treatment, during shipment and prior to manufacture such that the product performance can be optimised.

4 CONCLUSIONS

The application of broad spectrum biocides to control mould and sapstain fungi on freshly felled timber has been shown to be a complex process involving a wide range of physical, chemical and biological factors. The effects of poor post-treatment handling have been shown to be as important as chemical formulation and stability. Understanding the interaction between biocide, wood substrate and the micro-organisms involved is vital to the development of effective antisapstain compounds and wood preservatives.

REFERENCES

1. G.R. Williams, IRG 1990, IRG/WP/1437.
2. D.J. Dickinson, Int. Biodet. Bull., 1988, 24, 321.
3. G.R. Williams and R.A. Eaton, Proc. Int. Union For. Res. Org. 1987

4. G.R. Williams, 'Sapstain and Mould Control of Freshly Felled Timber', PhD Thesis, Portsmouth Polytechnic.
5. M.L. Edlund and B. Henningsson, IRG 1982, IRG/WP/3205.
6. D.A. Lewis, G.R. Williams and R.A. Eaton, Rec. Ann. BWPA, 1985, 14.
7. J. Mansen, IRG 1984, IRG/WP/3295.
8. S. Milano, IRG 1981, IRG/WP/3169.
9. D.A. Lewis and M. Spence, Proc. Int. Union For. Res. Org., 1985, S5.03
10. C.R. Rose, Proc. Int. Union For. Res. Org. 1987, 345
11. P.A. Briscoe, G.R. Williams, D.G. Anderson and G.M. Gadd, IRG 1990, IRG/WP/1464.
12. R.J. Wallance, R.A. Eaton and G.R. Williams, 1990, 'The Effect of Micro-organisms on the Biocidal Efficacy of the Antisapstain Compound Methylene-bis-thiocyanate' (in print) 'Biodeterioration 8' Elsevier Press.

Some Applications of Boron and Zinc Organic Compounds in Timber Preservation

David J. Love

RHÔNE-POULENC CHEMICALS LIMITED, ASHTON NEW ROAD, MANCHESTER M11 4AT, UK

1 INTRODUCTION

In recent years boron and zinc compounds have enjoyed increasing application as wood preservative biocides. This has been due to their effectiveness against a wide range of wood destroying organisms and their relatively low mammalian toxicity and limited impact on the environment. This paper describes some recent developments in the use of boron and zinc organic compounds for the preservation of timber.

2 SOLID BORON ESTERS

Inorganic boron compounds are well established as preservatives for timber and other materials against attack by fungi or insects. Recent developments in this field have included the use of shaped rods, pellets or tablets of inorganic boron compounds for insertion into pre-formed cavities in timber. However, preservative elements made from inorganic boron compounds suffer from several limitations. Due to high melting points (around 800°C) they are difficult and expensive to manufacture, prone to brittleness and may not be effective at low levels of moisture.

An alternative is the use of preservative rods made from hydrolysable solid boron esters. Such compounds having fairly low melting points (<200°C) can be formed into shaped elements more easily than is the case for inorganic boron compounds alone. In service the boron esters liberate boric acid due to moisture ingress and diffusion occurs in the presence of quite low levels of moisture. The use of boron esters in this way is disclosed in a recent patent[1].

A thorough investigation has been undertaken into the preparation of boron ester based rods, with and without inorganic boron compounds, and their diffusion characteristics in timber at varying moisture levels determined. The results of these investigations lead to the following conclusions:

- boron ester rods can readily be made and inorganic borates added if required
- a wide range of diffusion rates is possible
- satisfactory diffusion rates are achievable at low moisture levels
- addition of inorganic boron compounds modify diffusion rates
- boron concentrations in timber after diffusion are sufficient to provide biological control

Preparation of Esters. A number of solid esters of boric acid were prepared from a variety of hydroxylic compounds, including alcohols, glycols and an alkanolamine. The compounds varied with respect to melting point, hydrolytic stability and effective boric acid content (Table 1).

Preservative Rods. Rods were prepared by heating the esters to above their melting points and pouring into a steel mould. The mould was cooled to room temperature and broken to release the cast rod of dimensions 10.5 x 1.2 cm.

In some examples the effective boric acid content was boosted by the addition of boric acid (H_3BO_3) or disodium octaborate tetrahydrate ($Na_2B_8O_{13}.4H_2O$).

For the purpose of comparison commercial rods based on inorganic boron compounds were included in the tests.

Timber Samples. Specimens of joinery redwood (pine) of dimensions 450 mm x 94 mm x 44 mm were used for the diffusion trials. These were drilled at the mid point through the 94 mm dimension to within 6 mm of the bottom surface.

The preservative rods were cut to length and inserted into the holes with a sufficient upper gap to take a rubber bung (Fig.2).

Sample	Alcohol Type	% Boric Acid	Casting Temperature*
A	2, 2-dimethyl-1, 3 propanediol	34.9	180
A1	A + Boric oxide	120.7	140
B	2, 3-dimethyl-2, 3 butanediol	33.7	180
C	2, 6 dimethyl-4-heptanol	13.7	90
D	2-methyl-2, 4-pentanediol	41.2	75
D1	D + Boric acid	70.9	68
D2	D + Boric acid	70.9	150
D3	D + Boric acid	80.1	80
D4	D + $Na_2B_8O_{13}.4H_2O$	76.6	75
E	Alkanolamine	30.9	140
F	Ethylene glycol	61.2	130
G	Ethylene glycol	75.5	150
H	Inorganic 1+	92.6	-
I	Inorganic 2+	145.8	-
J	Ethylene glycol	59.5	130
J1	J + $Na_2B_8O_{13}.4H_2O$	65.79	155
K	Ethylene glycol	76.6	150
K1	K + $Na_2B_8O_{13}.4H_2O$	77.2	137
K2	K + $Na_2B_8O_{13}.4H_2O$	80.1	175
L	2-methyl-2, 4-pentanediol	38.9	70
L1	L + $Na_2B_8O_{13}.4H_2O$	60.1	70
L2	L + $Na_2B_8O_{13}.4H_2O$	79.5	85

* Casting temperatures could be some 10 to 20°C above melting point.

\+ Commercially available rods based on inorganic borates.

Table 1. Composition of boron esters

Equipment. Two thermostatically controlled forced air circulation cabinets were used to provide atmospheres of known relative humidity.

Trays containing saturated ammonium nitrate solution were placed in one cabinet to produce a relative humidity of 65%. To obtain 100% in the other cabinet, trays of water were used. The temperature of both cabinets was controlled at 23 ± 1°C.

Diffusion Trials. Sufficient numbers of wood samples had been prepared for examination at monthly intervals over a total period of six months.

The timber samples were stacked in the cabinets in layers with wooden spacers to allow air circulation.

The boron esters react with water (Fig.1) to form boric acid and glycol (or alcohol). The rate of hydrolysis depends upon the stability of the particular compound and the concentration of moisture in the timber. Diffusion of the boric acid throughout the timber would be expected to be assisted by the presence of the hydroxylic compound.

For assessment of the extent of diffusion timber samples were cut into 25 mm sections at two monthly intervals as shown in Fig. 2. The surface of each of the sections was sprayed with curcumin test reagents. For the final assessment at the end of six months, approximately 3 mm thick sections were removed for analysis of boric acid content.

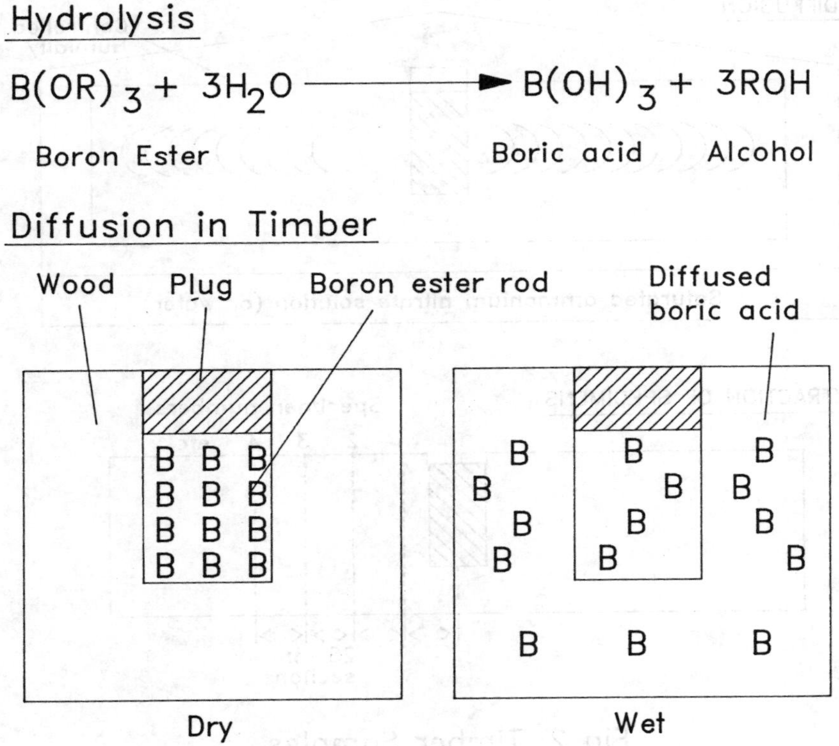

Hydrolysis

$$B(OR)_3 + 3H_2O \longrightarrow B(OH)_3 + 3ROH$$

Boron Ester Boric acid Alcohol

Diffusion in Timber

Wood Plug Boron ester rod Diffused boric acid

Dry Wet

Fig 1: Hydrolysis of boron esters

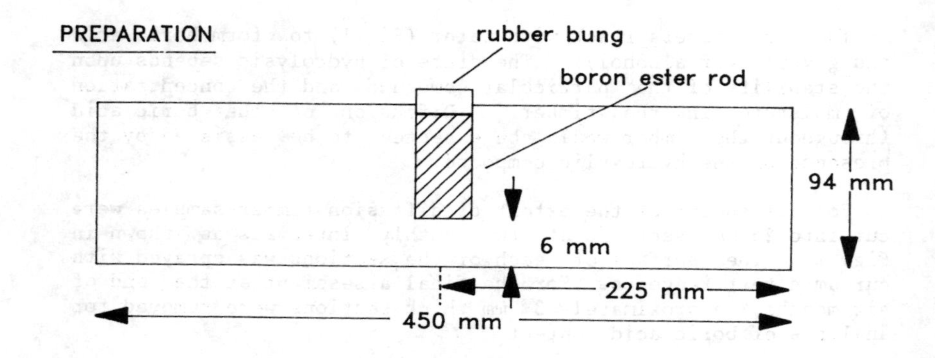

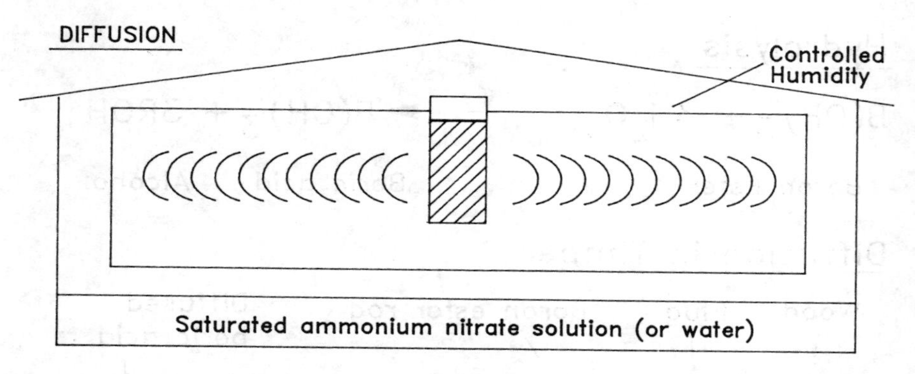

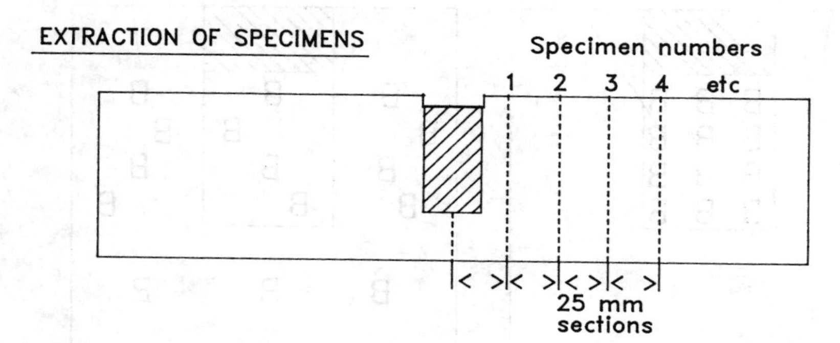

Fig 2: Timber Samples

Results and Conclusions. For the purpose of this work it was assumed that a minimum retention in redwood sapwood of 0.2% boric acid on oven dried wood would provide a practical level of protection against wood destroying organisms (see BWPDA Standard 105 for disodium octaborate tetrahydrate).

A red colour is produced by the curcumin in boron treated timber where the loading is 0.3% boric acid equivalent or more and this was used as an assessment of diffusion distance. The results from both the curcumin test and the final analysis for boric acid content are given in Table 2 and 3.

This investigation has shown that a wide range of diffusion rates is possible depending on the choice of ester or mixtures of esters, and in contrast to inorganic borates satisfactory diffusion rates in timber are possible even at moisture contents well below fibre saturation. Addition of inorganic boron compounds modifies the diffusion rates and would be expected to improve cost effectiveness.

Levels of boric acid achievable after diffusion in moist timber would be expected to provide protection against wood destroying organisms.

3 ZINC 2-ETHYL HEXANOATE

Zinc 2-ethyl hexanoate is an effective wood preservative but has a low mammalian toxicity, thus making it an ideal constituent for remedial and DIY wood treatments. Solvent-free it contains 22% zinc but for convenience it can be supplied in white spirit solution for example at 12% metal content. The typical properties of these two forms of zinc 2-ethyl hexanoate are shown below.

	% Metal Content	Colour	Specific Gravity at 25°C	Visc. Poise at 25°C	NVC (w/w)
Zinc 2-ethyl hexanoate 22%	22.0 ± 0.4	Pale Yellow	1.16	200	100%
Zinc 2-ethyl hexanoate 12%	12.0 ± 0.2	Pale Yellow	0.96	1	55%

Sample	Diffusion Distance (mm) 2 Months	4 Months	6 Months	Boric Acid Content After 6 Months (% on dry weight) Distance From Rod (mm) 25	50	75	100
A	75	100	100	0.67	0.20	0.10	0.03
A1	<25	25	25	0.06	<0.01	<0.01	<0.01
B	50	50	75	0.50	0.09	0.02	0.01
C	<25	<25	25	0.01	0.01	<0.01	<0.01
D	<25	25	25	0.01	<0.01	<0.01	<0.01
D1	<25	25	25	0.04	0.01	<0.01	<0.01
D2	<25	25	25	0.08	0.02	0.02	0.01
D3	<25	25	25	0.02	<0.01	<0.01	<0.01
D4	<25	25	25	0.08	0.01	<0.01	<0.01
E	-	50	50	0.09	0.02	0.01	<0.01
H	<25	<25	<25	<0.01	<0.01	<0.01	<0.01
I	<25	<25	<25	<0.01	<0.01	<0.01	<0.01
J	25	50	25	0.027	0.015	0.01	0.004
J1	25	50	25	0.01	0.008	0.007	0.005
K	50	50	50	0.062	0.017	0.01	0.01
K1	25	50	50	0.078	0.023	0.01	0.008
K2	25	50	25	0.03	0.02	0.014	0.014
L	-	25	-	0.008	0.08	0.007	0.007
L1	-	25	-	0.01	0.006	0.007	0.006
L2	-	25	-	0.013	0.01	0.008	0.009

Table 2: Boron Diffusion - Timber at 65% RH (12% Moisture)

Sample	Diffusion Distance (mm) 2 Months	4 Months	6 Months	Boric Acid Content After 6 Months (% on dry weight) Distance From Rod (mm) 25	50	75	100
A	75	100	100	0.59	0.17	0.06	0.02
A1	<25	50	50	0.22	0.02	0.01	0.01
B	75	75	75	0.41	0.13	0.03	0.01
C	<25	<25	25	0.02	0.01	<0.01	<0.01
D	25	50	50	0.47	0.12	0.01	<0.01
D1	25	50	50	0.58	0.12	0.02	<0.01
D2	25	50	50	0.36	0.06	0.01	<0.01
D3	25	50	50	0.49	0.10	0.02	0.01
D4	25	50	50	0.36	0.05	<0.01	<0.01
E	-	50	50	0.92	0.23	0.01	<0.01
H	25	25	25	0.05	<0.01	<0.01	<0.01
I	<25	25	25	0.06	<0.01	<0.01	<0.01
J	150	125	75	1.58	0.457	0.178	0.062
J1	75	150	75	2.33	1.57	0.546	0.186
K	125	150	50	1.74	0.82	0.067	0.054
K1	150	125	50	0.853	0.325	0.254	0.290
K2	100	125	50	1.023	0.324	0.358	0.203
L	25	50	25	0.304	0.098	0.048	0.041
L1	25	50	100	1.31	0.475	0.064	0.045
L2	25	50	25	0.704	0.189	0.051	0.045

Table 3: Boron Diffusion - Timber at 100% RH (22% Moisture)

Toxicity. The acute median lethal oral doses (LD_{50}) and their 95% confidence limits to rats for the 100% zinc 2-ethyl hexanoate have been estimated to be:

Males and females combined	6.6 (5.8 to 7.4) g/kg body weight
Males only	7.5 (6.4 to 8.8) g/kg body weight
Females only	5.9 (5.0 to 6.8) g/kg body weight

The low mammalian toxicity of zinc soaps enables them to be used in places inhabited by bats. Under UK law it is an offence to harm or disturb bats, or damage their roosts. It has been reported by Racey and Swift[2] of the University of Aberdeen that timber treated with fluids containing chemicals such as gamma-HCH and pentachlorophenol can be lethal to roosting bats. In the same paper the authors also reported that no obvious harm was caused to bats roosting in contact with timber treated with fluids containing zinc 2-ethyl hexanoate or esters of boron.

Application. Zinc 2-ethyl hexanoate finds wide use for in-situ applications such as DIY and horticultural applications and is normally diluted in a volatile organic solvent, usually white spirit, to be applied by brush, spray or dipping.

Solutions prepared from appropriate solvents do not change the dimensions or shape of timber nor do they corrode brackets or other metal parts which may be attached. Timber to be preserved should be reasonably dry before treatment. Any paint varnish or other surface coatings should be removed. Any inner or outer barks should also be removed as otherwise penetration of the preservative may be negligible.

Zinc 2-ethyl hexanoates are effective against fungi and wood boring insects and are particularly recommended for internal woodwork especially where the green colour of copper naphthenate or the high toxicity of pentachlorophenol may be undesirable. French polishes or other surface finishes can be applied to furniture and woodwork treated with Manosec Zinc 2-ethyl hexanoate once the solvent has fully evaporated.

With Pentachlorophenol. Zinc 2-ethyl hexanoate can be used in conjunction with pentachlorophenol which will increase the protective value of formulations. The pentachlorophenol will combine with zinc octate to form zinc pentachlorophenates and the volatility and rate of leaching of the pentachlorophenol will consequently be greatly reduced, whilst the biocidal activities of both the pentachlorophenol and the metal will be preserved.

Contact Insecticides. In certain applications, such as the treatment of a badly infected timber, contact insecticides such as gamma-hexachlorocyclohexane (known commercially as gamma HCH) or Pyrethroids such as Permethrin or Cypermethrin can also be added to zinc 2-ethylhexanoate formulations.

Zinc 2-Ethyl Hexanoate versus Zinc Naphthenate. Zinc naphthenate has been used for many years as a wood preservative and it has been known for some time that the various other zinc carboxylates, including zinc 2-ethylhexanoate, will give comparable protection.

Tests on small blocks of *Pinus sylvestris* (Scots pine) carried out at the Penarth Research Centre[3] indicated the following toxic limits for protection against Coniophora puteana:

	% w/w Zinc in Treating solution	Kg Zinc/m^3 wood
Zinc 2-ethyl hexanoate	0.4 - 0.63	1.95 - 3.10
Zinc Naphthenate	0.63 - 1.00	2.80 - 4.50

The 2-ethyl hexanoic acid used in the manufacture of the zinc is much paler and much less variable in composition compared with naphthenic acids. Hence zinc 2-ethyl hexanoate is much paler and more consistent in quality than zinc naphthenates. Solutions of zinc 2-ethyl hexanoate also have much lower viscosity than the corresponding zinc naphthenate solution at the same zinc content. This is expected to lead to an increased rate of penetration into timber. The following table gives the viscosity in centipoise of zinc 2-ethyl hexanoate and zinc naphthenate solution at several zinc concentrations in white spirit:

	3% Zinc	6% Zinc	12% Zinc
Zinc 2-ethyl hexanoate	1	1.5	100
Zinc naphthenate	22	340	660

When replacing zinc naphthenate by zinc 2-ethyl hexanoate it is desirable that substitution is made in terms of the amount of metal e.g. if a zinc naphthenate solution containing 3% zinc has hitherto been used then it should be replaced by a zinc 2-ethyl hexanoate solution also containing 3% zinc.

Blue Stain Control. Tests on Pinus sylvestris (Scots pine) sapwood at the Penarth Research Centre[4] have demonstrated that useful control of blue stain can be achieved by treating the timber with a zinc 2-ethyl hexanoate solution containing 3% zinc. Protection similar to that achieved by 1 or 2% solutions of dichlofluanid and better than that shown by a zinc naphthenate solution at the same zinc concentration was observed.

The blocks (dimensions 110mm x 40mm x 10mm) were assessed for the extent of surface stain at monthly intervals. The assessment of stain is based on the following scale:-

0 - not blue stained
1 - sparingly stained
2 - moderately stained in spots and streaks
3 - blue stained continuously to a maximum of one third or in streaks to one half
4 - strongly blue stained; more so than category 3

The monthly stain assessments are given in the table below, the assessments of stains for the 4 blocks are totalled so that the maximum stain rating possible is 16.

The depths of the stain free zones are also given in the table. Six measurements were made on each block which gives a total of twenty four measurements per toxicant. For clarity, all the measurements are not reported instead average values for the stain free zones at the upper (U) and lower (L) ends of the middle section are given. In general, the stain free zone at the lower end will be less than that at the upper end due to the moisture gradient set up in the block.

After the twelve month assessment, the blocks were cross-cut to give three equal sized blocks 35 mm in length. The depth of the stain free zone on each of the cross-cut surfaces of the middle block (the face nearest to the lower end is reference L and that of the upper end is reference U) was measured at three places - at the middle and at a distance of ten millimetres on each side of the middle.

TOXICANT	1	2	3	4	5	6	7	8	9	10	11	12	Stain free zone U	L
Untreated	0	3	5	6	6	6	12	16	16	16	16	16	0	0
2% dichlofluanid	0	0	0	0	0	0	0	1	4	4	4	4	9	9
1% dichlofluanid	0	0	0	0	0	0	0	0	4	4	4	7	8	8
Zinc 2-ethyl hexanoate (3% zinc)	0	0	0	0	0	0	4	4	4	4	4	5	8	8
Zinc naphthenate (3% zinc)	0	0	0	2	3	5	4	4	8	8	8	8	7	5

4 ZINC CARBOXYLATES FOR PRETREATMENT

Zinc compounds such as the naphthenates and more recently zinc 2-ethyl hexanoate have become well established for in-situ applications. For pretreatment applications these zinc compounds were thought unsuitable as sole biocides due to their propensity to leaching. With increasing environmental concern regarding the more toxic biocides such as tributyl tin oxide, recent developments have sought to produce zinc carboxylates with improved leaching resistance. This has been achieved by the use of new synthetic carboxylic acids such as neodecanoic acid. Wood preservatives suitable for joinery pretreatment and based on zinc salts of these acids are now appearing in the HSE Pesticides listing and it is expected that they will assume greater importance in the years to come.

REFERENCES

1. Rhône-Poulenc Chemicals, US Patent 4,911,988 March 1990.
2. Racey, P.A. and Swift, S.M., The Residual Effects of Remedial Timber Treatments on Bats, Biological Conservation 35 (1981) 205-214.
3. Penarth Research Centre, Report of private study carried out on behalf of Manchem Limited, May 1979.
4. Penarth Research Centre, Report of private study carried out on behalf of Manchem Limited, August 1979.

DJL/JMS/2973
23.5.91

Subterranean Termite Control in Buildings†

J.W.M. Logan

NATURAL RESOURCES INSTITUTE, CENTRAL AVENUE, CHATHAM MARITIME, CHATHAM, KENT ME4 4TB, UK

D.S. Buckley

44 ST RUALDS CLOSE, WALLINGFORD, OXFORDSHIRE OX10 0XE, UK

1 TERMITE DAMAGE TO BUILDINGS

There is little doubt that in some temperate and most tropical parts of the world, termites pose by far the greatest threat to unprotected timber used for construction and public amenity purposes. Estimates of the total cost of controlling termite damage are difficult to obtain but Edwards and Mill (1986) estimated that the world-wide cost of treating building timber damaged by termites in 1986 was nearly $2000 million US. Termite damage in the USA alone was estimated to cost property owners $750 million - $1.5 billion (Anon, 1984; Su & Scheffrahn, 1990) and, in 1986 the market for termiticides was worth $50 million (Hall, 1986). Consequently, the loss of the organochlorine (OC) insecticides, which were the mainstay of the termite control industry for over 30 years, has had major repercussions.

Termite damage to structures is caused either by drywood termites, Kalotermitidae, which nest within the timber and have no contact with soil, or by subterranean termites which nest in soil or rotten wood and forage into buildings and attack unprotected timber. This paper will address the methods of control for the latter which

------

†This article first appeared in *Pesticide Outlook*, (1991) **2**(1) 33-37 and is reprinted here with the permission of the Editor.

cause about 95% of termite damage to constructional wood in the USA.

Only a few of the many species of subterranean termite cause significant damage to buildings. These are *Reticulitermes* spp. (USA, southern Europe, China and Japan), *Coptotermes* spp. (USA (*Coptotermes formosanus* only), South America, Australia, and Asia), *Heterotermes* spp. (southern USA and Asia), *Nasutitermes* spp. (South America and Australia), *Psammotermes* spp. (Africa and Near East) and some *Macrotermitinae* (Africa and Asia). For a more detailed list, see Harris (1971).

2 TERMITICIDES PRE-1980

For many years pest control industries around the world have relied on chemicals to control termite damage. Before the Second World War, termite control depended on chemicals such as sodium arsenate and sodium fluosilicate to treat soil under buildings, fumigants (including arsenic salts and carbon tetrachloride) and dusts such as borax, Paris green and arsenical smelter dust (Kofoid et al., 1934). The development of the cyclodiene OC's, aldrin, chlordane, dieldrin and heptachlor, revolutionised the industry. These insecticides were cheap to produce and total control could be achieved for up to 30 years (Mauldin et al., 1987). Unfortunately, in the 1970s their long persistence combined with undesirable side effects was to form the basis for strong lobbying by environmentalists for the removal of these insecticides from the market in the USA and many other countries.

Use of these OC's in agriculture, where they could be taken up in food chains or washed into waterbodies, was shown to have certain detrimental effects on the environment, but, properly applied under houses in formulations with low mobility in the soil, they were relatively harmless and continued to be used until the 1980s. However, in 1985, following a wave of bad publicity related to alleged misuse and indications that these termiticides induced cancers in mice (Cooney 1984), Aldrin 4E was withdrawn from the USA market by the manufacturer (Anon, 1985a) and New York State banned the use of chlordane, aldrin and dieldrin (Anon, 1985b). Chlordane continued to be used in the USA until it was withdrawn and then banned in 1988 (Mampe, 1989).

The manufacturers of chlordane, Velsicol Chemical Corporation, still contend that chlordane is safe when properly applied. They claim it is not carcinogenic and cite health studies involving thousands of pest control operators and manufacturing personnel which show no increased levels of cancer (Cooney, 1984; Dysart, 1990). They have until 1992 to submit new data to the United States Environmental Protection Agency (USEPA) but, in view of the continued pressure from environmental groups, it is unlikely that chlordane will be reinstated.

Cyclodiene OC's have been the cornerstone of the termite control industry for close to 4 decades; more than 30 million homes were treated with chlordane in the USA between the early 1930s and mid 1980s (Hall, 1986). Despite the US ban, OC's continue to be used in some parts of the world (Fowler & Forte, 1990; Lenzet et al., 1990; Vongkaluang, 1990).

3 ALTERNATIVES TO ORGANOCHLORINE INSECTICIDES

Few people could have foreseen the sudden and dramatic swing away from OC's. The net result was to force the pace on the search for novel chemicals for termite control and on those that were already undergoing development. However, the high standards of control set by the OC's were to prove difficult to emulate. The primary requirements for a chemical to be used for termite control in buildings are:

1) It should be active against termites.

2) It should have low mammalian toxicity to protect applicators and residents.

3) It should have low water solubility to prevent leaching into ground water.

4) It should have low mobility in soil to avoid contamination of adjacent soil.

5) It must remain effective for a reasonable length of time, i.e. more than 5 years.

6) It must be accepted by the customers; they must be convinced it is safe. Customers now place considerable emphasis on low volatility and absence of odour.

7) It should be relatively inexpensive to manufacture and easily formulated.

In most countries, candidate termiticides undergo stringent evaluation before they can be registered for use. Around 7 years are required for the complete development process culminating in the marketable product. In the USA, they are tested against pest termite species in the laboratory and promising samples are retested at 6 month intervals to assess persistence. Those still effective after 2 years are sent for field testing to locations that represent semi-arid (Arizona), temperate (Maryland, Mississippi and South Carolina), sub-tropical (Florida and Midway Atoll) and tropical (Panama) climates. In these locations OC termiticides have provided 100% protection for more than 20-30 years. It is unlikely that any new insecticide will provide such protection. The USEPA now generally requires, in addition to stringent tests on mammalian toxicity and environmental acceptability, field data which show that a potential termiticide can remain 100% effective for at least 5 years at 3 test sites, or 7-10 years at 80-100% effectiveness (Mauldin et al., 1987).

Two field test methods are employed; the ground board and the concrete slab. In ground board tests, an untreated sapwood pine board is placed on top of the treated area of soil. For the concrete slab method, a wooden bait is fitted into the centre of a concrete slab laid over treated soil to simulate the base slab of a house. In both cases the termites must tunnel through the treated soil to attack the wood. The board and bait are inspected at regular intervals to assess termite damage. Full descriptions of the tests are given in Kard et al. (1989) and Mauldin et al. (1987).

Results of trials using some of these termiticides which are currently under test are summarised in Table 1.

4 ORGANOCHLORINES

Although use of these chemicals is now in decline, some have remained under test. As expected, they gave complete protection for a long time when applied at recommended concentrations. In ground board tests 0.125% chlordane was 100% effective for at least 5 years in

Arizona, Florida, Mississippi and South Carolina. In Maryland 100% efficacy was retained for 15 years and 80% for 22 years. At 0.5% and 1.0% total control was maintained for more than 19 years at most sites. Groundboard tests using a 2:1 chlordane + heptachlor mixture remained 100% effective for 28 years at concentrations ranging from 0.25-1.0%.

Table 1 Termiticides Currently Undergoing Evaluation In USA (from Kard et al., 1989).

Common Name	Trade Name	Manufacturer
ORGANOCHLORINE		
Chlordane	Gold Crest C-100	Velsicol
Heptachlor	Gold Crest H-60	Velsicol
ORGANOPHOSPHORUS		
Chlorpyrifos	Dursban 1C	Dow Elanco
Isofenphos	Pryfon 6	Mobay
PYRETHROID		
Bifenthrin	Capture	FMC
Cyahlothrin	Karate	ICI Americas
Cyfluthrin	Tempo	Mobay
Cypermethrin	Demon TC	ICI Americas
Esfenvalerate	Asana	Du Pont
Fenpropathrin	Danitol	Valent USA
Permethrin	Torpedo	ICI Americas
Permethrin	Dragnet	FMC
S276B	-	Roussel Bio
Tralomethrin	Scout	Roussel Bio

5 SYNTHETIC PYRETHROIDS

These chemicals are noted for their high insecticidal activity and low mammalian toxicity. Permethrin at 0.5% concentration remained 100% effective in concrete slab tests for more than 10 years in Arizona and for 5 years in Mississippi and South Carolina, dropping to 90% efficacy in the 6th year. One per cent gave complete control for more than 10 years in Arizona, Florida and South Carolina. It performed badly in ground board tests in South Carolina and Mississippi but, at 1.0%, achieved 100% control for 9 and 6 years in Arizona and California, respectively. Cypermethrin has so far (6 years) remained 100% effective at concentrations of 1.0% at all test locations both in concrete slab tests and ground board tests (which are still continuing). Fenvalerate (1.0%) was 100% effective at all test locations for 6 years in concrete slab tests, and for at least 4 years in ground board tests (Kard et al., 1989; Mauldin et al., 1987). The pyrethroid insecticides have the additional advantage that they are repellent to termites(Su et al., 1982).

6 ORGANOPHOSPHORUS (OP) INSECTICIDES

Concrete slab tests using 1.0 and 2.0% chlorpyrifos have remained 100% effective for more than 10 years (in one case 19 years) but in Florida and Arizona the lower concentration was less effective. Isofenphos remained 100% effective for at least 5 years at 0.5, 1.0 and 2.0% concentrations at Arizona, Florida and Mississippi but did not perform as well in ground board tests. Interestingly, none of the OP's performed well under hot, dry conditions (Kard et al., 1989; Mauldin et al., 1987). Similar results were found in Australian field tests (Lenz et al., 1990).

7 COMMERCIAL TERMITICIDES

Based on the results of these trials, certain products containing chlorpyrifos, isofenphos, permethrin, cypermethrin and fenvalerate were registered for use in termite control and are now used commercially in the USA. Endosulfan also showed sufficient promise to be registered but was never marketed for termite control. Although endosulfan is an OC, it is broken down rapidly by sunlight and so does not cause the environmental problems generally associated with these compounds.

Protected from light, it remained effective for more than 10 years and 5 years under concrete slabs and ground boards, respectively, at 0.5-2.0% at all test sites (Mauldin et al., 1987). Possibly the manufacturers feel that the OC connection would prejudice customer acceptability.

8 CURRENT TERMITICIDE USE IN USA

The withdrawal of the OC's has resulted in great changes in the termiticide market. In 1985, the bulk of the pest control operators (PCOs) in the USA used chlorpyrifos (61.4%), chlordane + heptachlor mixture (52.4%) and/or chlordane (48.1%) (percentages add up to more than 100% because of multiple responses). Heptachlor (10.1%) and aldrin(4.8%) were also used and permethrin accounted for less than 4% (Anon,1986). By 1988 chlorpyrifos was used by 80.3% of PCOs, isofenphosby 16.4% and permethrin as *Dragnet*, by 10.8%, and as *Torpedo*,by 14.6%. Fenvalerate increased from 0.5% the previous year to 24.9% and cypermethrin appeared for the first time, used by 29.1% of PCOs (Mix, 1989).

9 ALTERNATIVE CONTROL METHODS

Biological Control

There has been considerable research into termite control using pathogens such as bacteria, fungi and nematodes. Despite promising laboratory experimental results, field trials have been disappointing with only short-term reductions in termite populations, possibly because the termites walled off sick and dying individuals from the rest of the colony and so prevented spread of infection (Logan et al.,1990), or because they possessed an ability to detect and avoid entomogenous nematodes (Epsky & Capinera, 1988; Mauldin & Beal, 1989). In spite of this, the nematode *Steinernema feltiae* (= *Neoaplectana carpocapsae*) is marketed as *Spear* in the USA for termite control. There is a strong interest in the potential for biological control and research is continuing.

Resistant Timber

Studies throughout the world have identified a wide range of termite resistant woods (Harris, 1971). If these woods are used throughout the building there is less need for chemical barriers although other items, such as books and clothing, may still be attacked, particularly in the tropics. Timber resistance depends either on hardness or on chemicals in the wood which repel or are toxic to termites. These chemicals can be extracted from poor quality resistant woods and used to protect high quality susceptible timber (Anon, 1984).

Physical Barriers

Well constructed concrete foundations, seals round cables and pipes and the use of metal termite shields can considerably reduce termite attack on buildings (Edwards & Mill, 1986). Protection can also be provided by a bed of sand with interstitial spaces too small to allow termites through but grains too large for the termites to move (Tamashiro et al., 1987).

10 INSECTICIDAL BAITS

There is a growing interest in the use of baits impregnated with pesticide to eliminate termite colonies around and within buildings. Potential control agents under investigation for use in baits include slow acting stomach poisons, insect growth regulators, protozoacides (to kill the termites' symbiotic protozoa) and fungicides (to control the fungus dependent Macrotermitinae) (Anon, 1984; El Bakri et al., 1989; Jones, 1987; Logan & Abood, 1990; Paton & Miller, 1980; Su et al., 1987). It is estimated that insecticidal baits would use 1000 fold less insecticide than current soil termiticide treatments (Su & Scheffrahn, 1990).

11 THE FUTURE FOR TERMITE CONTROL

Householders and PCOs are having to change their attitudes to termite control. The OC's were extremely effective in a wide range of climates and different soils and gave protection for more than 20 years; the new insecticides are shorter lived and performance varies with soil type and climate. The OP's persist longer in temperate areas while the pyrethroids fare particularly well in hot dry climates. Neither last as well as the OC's and retreatments for termite control every 5-10 years will become necessary in some areas. To facilitate retreatment, builders are now incorporating a network of tubing into the foundations of some new buildings during construction. Termiticide treatment can be applied from a single service point, reducing the disturbance and risk to the householder (Anon, 1990).

Despite the long field trials, these new insecticides have only been used commercially for a relatively short time. Problems associated with different soils and variations in application techniques are already appearing (Mampe, 1989). However, the choice of insecticides with different properties not only provides PCOs with alternative treatments but also encourages research and development of new formulations through competition between manufacturers. With close collaboration between the PCOs and the pesticide industry, carefully watched by a critical public, there are strong pressures to develop safe, reliable termite control treatments to suit every situation. It seems unlikely that there will be any serious alternatives to the use of chemicals for termite control in the near future. However, continuing pressure against the use of chemicals may result in greater use of physical barriers and termite resistant timber.

ACKNOWLEDGEMENTS

Thanks are due to Dr T.G. Wood and Dr R.A. Johnson for their comments on the draft of this paper.

REFERENCES

Anon (1984) Gulfport searches for new termite technology. *Pest Control*, **53** (2), 20-24.

Anon (1985a) Aldrin 4E pulled from pest control industry. *Pest Control*, **54** (4), 12.

Anon (1985b) New York bans aldrin, chlordane, dieldrin. *Pest Control*, **54** (4), 16.

Anon (1986) Termiticides: winds of change blow. *Pest Control*, **55** (3), 22-23.

Anon (1990) Pesticide tubes-in-walls. *Pest Control*, **58** (9), 15.

Cooney, K.M. (1984) Industry forcibly presents its case during NY hearings. *Pest Control*, **53** (6), 24,26.

Dysart, J. (1990) Chlordane's future. *Pest Control*, **58** (3), 56.

Edwards, R.; Mill, A.E. (1986). *Termites In Buildings-Their Biology And Control*. The Rentokil Library, East Grinsted,UK, 261pp.

El Bakri, A.; Eldein, N.; Kambal, M.A.; Thomas, R.J.; Wood,T.G. (1989) Effect of fungicide impregnated wood on the viability of fungus combs and colonies of *Microtermes sp. nr. albopartitus* (Isoptera: Macrotermitinae). *Sociobiology*, **15**,175-180.

Epsky, N.D.; Capinera, J.L. (1988) Efficacy of the entomogenous nematode *Steinernema feltiae* against the subterranean termite *Reticulitermes tibialis* (Isoptera, Rhinotermitidae). *Journal of Economic Entomology*, **81**,1313-1317.

Fowler, H.G.; Forte, L.C. (1990) Status and properties of termite problems and control in Brazil. *Sociobiology*, **17**,45-56.

Hall, R. (1986) New termiticides are effective but less persistent. *Pest Control*, **55** (3), 24, 26, 30.

Harris, W.V. (1971) *Termites their recognition and control*,Longman, London, xiii + 186 pp.

Jones, S.C. (1987) Extended exposure of two insect growth regulators on *Coptotermes formosanus* Shiraki. pp 58-61 In Tamashiro,M.; Su, N.-Y. (eds.) *Biology and control of the Formosan subterranean termite*. Hawaii Institute of Tropical Agriculture and Human Resources,Honolulu, Hawaii.

Kofoid, C.A.; Light, S.F.; Horner, A.C.; Randall, M.; Herms,W.B.; Bowe, E.E. (eds.) (1934) *Termites and termite control*. University of California Press, Berkley, California. 734 pp.

Kard, B.M.; Mauldin, J.K.; Jones, S.C. (1989) The latest on termiticides. *Pest Control*, **57** (10), 58, 60, 68.

Lenz, M.; Watson, J.A.L.; Barrett, R.A.; Runko, S. (1990) The effectiveness of insecticidal soil barriers against subterranean termites in Australia. *Sociobiology*, **17**, 9-36.

Logan, J.W.M.; Abood, F. (1990) Laboratory trials on the toxicity of hydramethylnon (Amdro; AC 217,300) to *Reticulitermes santonensis* Feytaud (Isoptera: Rhinotermitidae) and *Microtermes lepidus* Sjöstedt Isoptera: Termitidae). *Bulletin of Entomological Research*, **80**, 19-26.

Logan, J.W.M.; Cowie, R.H.; Wood, T.G. (1990) Termite (Isoptera) control in agriculture and forestry by non-chemical methods: a review. *Bulletin of Entomological Research*, **80**, 309-330.

Mampe, C.D. (1989) What we've learned about the new termiticides. *Pest Control*, **57** (2), 60-62.

Mauldin, J.K.; Jones, S.; Beal, R.H. (1987) Viewing termiticides. *Pest Control*, **56** (10), 46-59.

Mauldin, J.K.; Beal, R.H. (1989) Entomogenous nematodes for control of subterranean termites, *Reticulitermes* spp. (Isoptera:Rhinotermitidae). *Journal of Economic Entomology*, **82**, 1638-1642.

Mix, J. (1989) Dursban TC again named top termiticide. *Pest Control*, **57** (2), 56-58.

Mix, J. (1990) A look into the future. *Pest Control*, **58** (3), 50, 54.

Paton, R; Miller, L.R. (1980) Control of *Mastotermes darwiniensis* Froggatt (Isoptera : Mastotermitidae) with mirex baits. *Australian Forestry Research*, **10**, 249-258.

Su, N.-Y.; Tamashiro, M.; Yates, J.R.; Haverty, M.I. (1982) Effect of behaviour on evaluation of insecticides for prevention of or remedial control of the Formosan subterranean termite. *Journal of Economic Entomology*, **75**, 188-193.

Su, N.-Y.; Tamashiro, M.; Haverty, M.I. (1987) Characterization of slow-acting insecticides for the remedial control of the Formosan subterranean termite (Isoptera: Rhinotermitidae). *Journal of Economic Entomology*, **80**, 1-4.

Su, N.-Y.; Scheffrahn, R.H. (1990) Economically important termites in the United States and their control. *Sociobiology*, **17**, 77-94.

Tamashiro, M.; Yates, J.B.; Ebesu, R.H. (1987) The Formosan subterranean termite in Hawaii: problems and control. pp 15-22 In Tamashiro, M.; Su, N.-Y. (eds.) *Biology and control of the Formosan subterranean termite*. Hawaii Institute of Tropical Agriculture and Human Resources, Honolulu, Hawaii.

Vongkaluang, C. (1990) A review of insecticides used for the prevention of subterranean termites in Thailand. *Sociobiology*, **17**, 95-101.

Nitrogen Assimilation and Transport as Target Processes for Inhibitors of Fungal Growth, and the Effects of α-Aminoisobutyric Acid on Wood Decay Fungi

S.C. Watkinson

DEPARTMENT OF PLANT SCIENCES UNIVERSITY OF OXFORD, SOUTH PARKS ROAD, OXFORD OX1 3RA, UK

1 INTRODUCTION

This paper describes a possible method of protecting timber by inhibiting fungal growth. The chemical used is a nonmetabolised amino acid, the one investigated and used in most tests so far is α-aminoisobutyric acid (AIB). The principle on which use of this compound is based is that wood decaying Basidiomycete fungi are metabolically adapted to using wood as their sole food source. Since nitrogen is scarce in wood these fungi depend on a fine balance between uptake of nitrogen compounds and expenditure of nitrogen in growth and in the synthesis of extracellular enzymes. This balance can be upset by feeding the fungus with a non-metabolised amino acid such as AIB, which it takes up by the same mechanism as a metabolisable amino acid. However, AIB is useless in nitrogen metabolism while displacing the fungus's food amino acid. The effects of AIB are to inhibit spread of the fungal mycelium with little or no effect on biomass - that is, it controls growth without being fungicidal. Its toxicity is relatively low and it is freely soluble in water. It is of proven value in remedial treatment of outbreaks of dry rot due to *Serpula lacrymans*, where it can be used to slow the spread of mycelium in buildings. It may have wider applications to fungal control as it has a wide spectrum of action and some preservative activity has been shown.

2 ASPECTS OF THE NUTRITIONAL ECOLOGY OF BASIDIOMYCETE TIMBER DECAY FUNGI WHICH ARE POSSIBLE TARGETS FOR CONTROL METHODS.

Basidiomycete fungi which attack timber to an economically significant extent include Serpula lacrymans, Armillaria mellea, Coriolus versicolor, Pleurotus ostreatus, Lentinus lepidens, Phlebia gigantea, Coriophora puteana, Stereum sanguinolentum and Gloeophyllum trabeum (1). All form large colonies in wood, made up of microscopic filaments, the hyphae, which elongate by addition of material at their tips. Food for this growth is absorbed from the wood, which the fungus degrades by means of secreting extracellular enzymes, principally cellulases, proteinases and - in white rot fungi - ligninase. The solubilised products of enzyme action are absorbed into the hyphae, and utilised in metabolism. They can be absorbed in one part of the colony and translocated to another part for metabolism. Translocation can, in some species such as Serpula lacrimans, extend over distances of several metres. Thus four separate physiological processes are necessary for growth of fungi in timber: enzymic solubilisation (2) of wood components, mainly cellulose and protein (3), active uptake into fungal hyphae of sugars and amino acids so produced (4, 5); translocation of nutrients in solution to the growing edge of the colony (6); and the synthesis there of new cellular material and enzymes (7). Each of these processes is essential for fungal growth and could present a target for chemical inhibition.

Nitrogen Ecology of Wood Decay Fungi

It is the nitrogen nutrition of timber decay Basidiomycetes which appears most vulnerable to chemical interference. All fungi require carbon and nitrogen compounds, but while carbon is abundant in wood, nitrogen is scarce. It is present as protein (3) embedded in water-insoluble cellulose cell walls impregnated with lignin (2). The C/N ratio can be as much as 500:1 (8). When wood is the only food source the fungus must solubilise cellulose to reach the embedded protein, and as this requires the N-utilising processes of hyphal elongation and synthesis of enzyme protein, there is a fine balance between nitrogen uptake and expenditure. Serpula lacrymans can concentrate nitrogen absorbed from wood to the extent that mycelium contains 37 mg N/g compared to an original concentration in the wood of 0.6 mg/g (9).

Colonies of wood decaying fungi cycle nitrogen, retrieving it from older parts of the colony and translocating it to regions of growth (10). The levels of N normal in wood have been shown to be limiting for fungal dry weight production and for cellulose degradation (9). Nitrogen compounds supplied to one part of a colony can release the limitation in cellulolysis at another part of the colony (9). When mycelium of S. lacrymans grows from one food base to another over a non-nutrient surface, it can transfer up to 80% of the total N from the first to the second, fresh, food base (11). Translocation of nitrogen occurs in the hyphae in the form of free soluble protein amino acids (12). There is an intrahyphal balance between protein synthesis and breakdown to amino acids which is regulated by nitrogen availability - under N starvation more protein is broken down and mobilised as free amino acid. This probably enables the mycelium to buffer its internal nitrogen levels as it grows over N rich and N poor substrata.

Serpula lacrymans, of all timber decay Basidiomycetes, is notoriously capable of mycelial extension from a wood food base such as a timber joist, over non nutrient surfaces such as masonry (13). This, with its dependence on wood as sole nutrient source, makes its nitrogen cycling mechanisms obvious targets for control measures. Therefore most work on the use of AIB has been with this fungus although a wide range of fungi has been screened for an inhibitory effect (1).

Effects of α-Aminoisobutyric Acid on the Physiology of Fungi

The structure of AIB is shown in Fig.1.

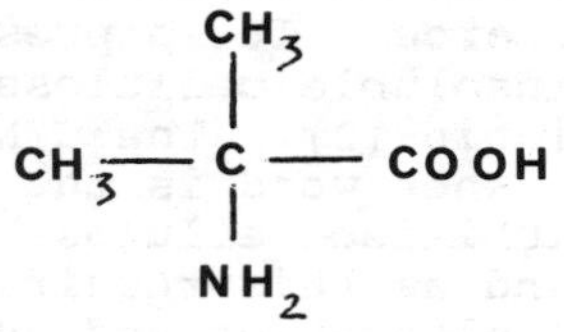

$$CH_3-\underset{\underset{NH_2}{|}}{\overset{\overset{CH_3}{|}}{C}}-COOH$$

It is a methyl-substituted analogue of alanine, forming odourless white crystals freely soluble in water. It is of relatively low toxicity, with LD_{50}, tested intraperitoneally on rats, of 750 mg/kg. Although AIB is not a protein amino acid it does occur naturally, having been found in some plants (14) and as a component of a peptide antibiotic (15). Alanine is a protein amino acid and a readily-utilised source of nitrogen for fungi which they would obtain in nature by hydrolysis of protein, or synthesise by transamination reactions in metabolism (16). Like other utilisable amino acids it is actively accumulated (18). It is broken down by deamination to pyruvic acid. When utilisable amino acids such as L-alanine are ^{14}C labelled and fed to fungi, the ^{14}C is either released as $^{14}CO_2$ or incorporated into products of cellular synthesis such as protein.

Like alanine, AIB is actively taken up by fungal hyphae via transport proteins located in the cell membrane, which show some specificity and selectivity for different amino acids (17). Uptake of AIB by hyphae of <u>Serpula</u> <u>lacrymans</u> is not inhibited by the presence in the culture medium of a utilisable amino acid, glutamic acid; however AIB does inhibit glutamic acid uptake, indicating that AIB is preferred to glutamic acid by the amino acid transport system (18).

Although AIB 'deceives' the transport system, once inside the cell it is not metabolised. Mycelium of <u>S.</u> <u>lacrymans</u> fed with ^{14}C-AIB does not release $^{14}CO_2$ or synthesise any radioactive insoluble material, although feeding with unlabelled AIB depresses synthesis of insoluble material from ^{14}C-glutamic acid (18). Chromatography showed ^{14}C-AIB to be the only labelled compound present in mycelial extracts 24 h after feeding, compared with ^{14}C-glutamate which releases $^{14}CO_2$ within minutes of feeding. Glutamate in the intra-cellular free amino acid pool was depressed to half its normal level by addition of 10% AIB (w/v) to the growth medium. This is interesting in relation to the use of AIB in control, as glutamic acid is a key intermediate in fungal nitrogen metabolism (19). It is also the amino acid produced most abundantly into the free amino acid pool under starvation-induced protein breakdown, and therefore presumably an important vehicle for mobilisation of N around the colony (12).

Once inside the hypha, AIB is translocated through the colony in the same way as utilisable amino acid (18). It is a useful marker for N movement as it is not metabolised en route.

The Effect of AIB on Fungal Growth and Development

Colonies of fungi respond to applications of AIB with diminished rates of spread, whether AIB is present throughout a growth medium (1), or applied locally to the food base from which mycelium is extending (11). Concentrations which inhibit extension - of the order of 1% AIB - have little effect on biomass. Microscopic examination shows that the form of the hyphae is altered, and instead of leading hyphae elongating at rates of approximately 3mm per day, their extension is replaced by an increased number of lateral branches, so that the colony becomes limited in extent, localised and bushy in appearance. The effect is durable with a single application continuing to be effective for eight months. Thus AIB is a paramorphogen - it controls fungal attack by altering development without killing the fungus or preventing biomass increase. Paramorphogentic action is a promising mode of control where low toxicity is needed (20). The cellular mechanisms by which development is altered are at present unknown, but merit further study as a target for control methods. There is evidence that the hypha can 'sense' intrahyphal amino acid levels and respond by elongation of hyphae to scavenge under N poor conditions, or localized branching to exploit a locally abundant food source (9), and amino acid analogues could subvert such a sensing system. Inhibition of extension by AIB has been shown in 16 out of 18 species of wood decay and wood inhabiting fungi tested (1). Most were affected at 1% AIB w/v. Inhibition was partially reversible in a concentration-dependent manner by glutamic acid or alanine (20).

Methods for the Control of Timber Decay Fungi using AIB and Other Amino Acid Analogues

Use of AIB is most appropriate when there is a need to control the spread of a mycelium which is using wood or other N-poor material as its sole nitrogen source. The chemical can be applied to the fungus rather than to the wood, and because the mycelium transports it, it can reach growing mycelium which is otherwise inaccessible to treatment. It seems particularly appropriate to remedial treatment of dry rot caused by S. lacrymans. Possible methods of application are as a water based spray, paste, or solution absorbed into cellulosic material such as cotton fabric on paper placed in contact with an accessible area of actively growing mycelium.

Preventive treatment would be possible by pretreatment of construction timber with AIB or the insertion of pellets to be activated by enzymes on water secreted by the fungus.

3 REFERENCES

1. M.L. Elliott and S.C. Watkinson, Int. Biodet., 1989, 25, 355.

2. R.A.P. Montgomery, 'Decomposer Basidiomycetes', Cambridge University Press, Cambridge, 1982, Chapter 3, p. 51.

3. R.A. Laidlaw and G.A. Smith, Holzforsch., 1965, 19, 129.

4. J. Whittaker, Trans. Brit. Mycol. Soc., 1965, 67, (3), 365.

5. D.H. Jennings, 'The Filamentous Fungi', Arnold, London, 1974, Vol. 2, Chapter 2, p. 32.

6. D.H. Jennings, 'Decomposer Basidiomycetes', Cambridge University Press, Cambridge, 1982, Chapter 5, p.91.

7. J.H. Burnett, 'Fungal walls and hyphal growth' Cambridge University Press, Cambridge, 1978. Chapter 1, p. 1.

8. W. Merrill and E.B. Cowling, Phytopathol., 1966, 56, 1083.

9. S.C. Watkinson, E.M. Davison and J. Bramah, New Phytol., 1981, 89, 295.

10. M.P. Levi and E.B. Cowling, Phytopathol., 59, 460.

11. S.C. Watkinson, 'Physiology and Ecology of the Fungal Mycelium', Cambridge University Press, Cambridge, 1984, Chapter 8, p. 165.

12. C.E. Venables and S.C. Watkinson, Mycol. Res., 1989, 94, 289.

13. J. Ramsbottom, 'Mushrooms and Toadstools', Collins, London, 1977, Chapter 20, p. 233.

14. S. Hunt, 'Chemistry and Biochemistry of the Amino Acids', Chapman and Hall, London, 1985, Chapter 4, p. 55.

15. R.C. Pandey, J.C. Cook and J. Reinhart, Am. Chem. Soc., 1978, 99, 8469.

16. J.A. Pateman and J.R. Kinghorn, 'The Filamentous Fungi', Ed. Arnold, London, Vol. 2, Chapter 7, p. 159.

17. S. Ogilvie-Villa, R. Mason deBusk, and A.G. deBusk, Bacteriol., 1981, 147, 944.

18. S.C. Watkinson, FEMS Lett., 1984, 24, 247.

19. A.P.J. Trinci, 'Mode of action of antifungal agents', Cambridge University Press, Cambridge, 1984, Chapter 6, p. 113.

20. M.L. Elliott, M.Sc Thesis, University of Oxford, 1987.

Subject Index